LIMINAL ZONES

LIMINAL ZONES
Where Lakes End and Rivers Begin

Kim Trevathan

THE UNIVERSITY OF TENNESSEE PRESS / KNOXVILLE

Copyright © 2013 by The University of Tennessee Press / Knoxville.
All Rights Reserved. Manufactured in the United States of America.
First Edition.

Some passages from this book have appeared, in different form, in the following: *New Madrid* (July 2012); *Still: The Magazine* (fall 2011); the *Maryville Daily Times* (10 July 2010, 25 July 2010, 31 July 2011); and *Metro Pulse* (11 Oct. 2007; 21 Oct. 2010; Oct. 13, 2011).

The paper in this book meets the requirements of American National Standards Institute / National Information Standards Organization specification Z39.48-1992 (Permanence of Paper). It contains 30 percent post-consumer waste and is certified by the Forest Stewardship Council.

Library of Congress Cataloging-in-Publication Data

Trevathan, Kim, 1958–
Liminal zones: where lakes end and rivers begin / Kim Trevathan.
      pages cm
ISBN 978-1-57233-953-8 (paperback) — ISBN 1-57233-953-5 (paperback)
1. United States—Description and travel.
2. Trevathan, Kim, 1958–Travel—United States.
3. Rivers—United States.
4. Dams—Environmental aspects—United States.
5. Canoes and canoeing—United States.
6. Limnology—United States.
I. Title.

E169.Z83T74 2013
973'.0946—dc23
2012040203

Thanks to my family, who support and encourage my writing.

Thanks to Maryville College for its support.

CONTENTS

Introduction   xi

Part I: A Season Bereft
1. The Big South Fork: Productive Failure   1
2. The Nantahala: The Liminal Unveiled   11
3. My History with Dams   21

Part II: Road Trip of Rivers
4. The Concept   33
5. Easy Water: The Tippecanoe and the James   39
6. The Rogue's Embrace   53
7. Aesthetic Convergence:
   The Clearwater and the Deschutes   67
8. Reconsidering the Liminal:
   The Dolores, the Conejos, and a Fractious
   Campground in Folsom, California   85

Part III: Brackish Waters
9. Big Lagoon to Maple Creek: From One World to Another   127
10. Fear, Delusion, and Peace on the Edisto   133

Part IV: Damaged Waters
11. Seeking Damaged Waters   145
12. Up Pistol Creek   149
13. Finding and Smelling the Pigeon   157

Part V: Night Paddling
14. Hematite   165
15. Energy   177

Part VI: Company
    16. With Libby on Hematite   187
    17. Navigating by the Stars up Citico Creek   193
    18. Warning: German Shepherd in Bow   199
    19. Final Thoughts   209

Epilogue: Letters   215
Bibliography   219

ILLUSTRATIONS

Figures

Following page 104

Transitional Zone between Fontana Lake and the Nantahala River
Gauley River/Summersville Lake Transitional Zone
Water Released from Shafer Lake in Northern Indiana
Hoosier on Shafer Lake
White River Outpost of the Bass Pro Shop
Negotiating the Rogue River
Salmon Lake
Campground at Salmon Lake State Park
Deschutes River
McPhee Reservoir
Platoro Reservoir
View from Campsite at Big Lagoon
River Otters
Cypress Trees Near Givhans State Park
Mouth of Pond Creek
Pistol Creek Graffiti
Pile of Deadfall and Trash on Pistol Creek
Spillway with Steps at Hematite Lake
Crossed Trees at Energy Lake
Niece Libby on Hematite Lake
Moon Above Citico Creek
Dark Creek
Norm on Quiet Creek
Mating Toads
Retrieving Ball
Norm Surveys Little Tennessee River

Maps

The Big South Fork of the Cumberland River   xvi
Nantahala and Fontana lakes   10
Shafer Lake/Tippecanoe River   38
Table Rock Lake/James River   44
The Rogue River   52
Salmon Lake/Clearwater River   66
Lake Billy Chinook/Deschutes River   77
Platoro Lake/Conejos River   98
Big Lagoon/Maple Creek   126
The Edisto River   132
Waterville Lake/Pigeon River   156
Land between the Lakes   164
Hematite Lake   186
Tellico Lake/Citico Creek   192

INTRODUCTION

I began a journey alone the summer after my traveling companion Jasper died. A German shepherd/lab mix who tilted his head back and forth when you talked to him, Jasper canoed with me thousands of miles, including 652 down the length of the Tennessee River—from Knoxville to Paducah. We locked through nine dams and endured blistering heat, drought, chill, thunderstorms, swampy campsites, and nauseating boat wakes. Five weeks it lasted, from late August into October. Though he vocalized his opinion continually, only one time on the trip did Jasper balk at taking his place in the bow. When I paddled away that morning, acting as if I were leaving him there at a snaky campsite in Mississippi, he sat on the bank and stared, calling my bluff until I'd drifted fifty yards away. Then he plunged into the tepid water in pursuit, swimming faster than I could paddle. That day we got hit with the worst storm of the trip, and by midmorning, when we were shivering together on the treeless riprap shoreline under the same poncho, I apologized for not listening to his advice to slow down. Seven years later my wife Julie and I carried Jasper out to the side yard next to the young tomato plants, where we put an end to the cancer that he had endured with such stoicism and dignity. He took the injection without a flinch, understanding, I think, that the slightest fidget would make us grieve harder.

    I grieved the same way I celebrated life with Jasper: by taking to the water, this time alone, a yellow kayak replacing the canoe. Instead of traveling from headwaters to mouth, I paddled upstream to the places where dead rivers, otherwise known as lakes or reservoirs, revive themselves against the influence of dams. These journeys, toward destinations I initially called transitional zones, could just as well have been called dying zones had I been approaching from the opposite direction, downstream toward the pooling of the reservoir, instead of upstream, away from dams. I preferred going upstream for the effort and for what the effort seemed to symbolize, an escape from human alterations of the landscape toward a region of purity, authenticity, and transcendence.

The word *liminal,* derived from the Latin word *limen,* for threshold, packed some of the metaphysical punch that I was seeking in these geographical quadrants. You can be in a liminal state of mind, in-between realities, so to speak, as you are in the process of waking up from a vivid dream, perhaps registering the bold-faced numbers of your alarm clock while babbling to some green, three-headed, tentacled nemesis floating at the foot of your bed. In anthropology a person undergoing a rite of passage, transitioning from one social state to another, from boy to man, for example, would be in a liminal state. Belden Lane, in his book *Landscapes of the Sacred,* has described it as "having left one place, one conventional state of being, and not yet having arrived at another . . . caught betwixt and between." In pop culture, it's embodied by *The Twilight Zone* (1959–64), the TV series that host Rod Serling introduced in his clipped narration, cigarette in hand, as "the middle ground between light and shadow, between science and superstition." Lane has described "liminal zones" as places that are "not-places . . . caught in transition, existing only on the margins of a structured world." In the natural world, according to Lane, the liminal might be represented by the mouth of a cave, a hedgerow, a swamp, the confluence of two rivers, a natural arch, or the edge of a cliff. For me it's where stilled waters begin to move.

I grew up near Kentucky Lake—part of the dammed-up Tennessee—and for decades I've enjoyed fishing, swimming, water-skiing, cruising, and paddling its flat water bays, tributaries, and main channel. Now I live in a land dominated by dammed rivers—East Tennessee—and I've canoed the length of the two main working rivers that rise in the region: the Cumberland and the Tennessee, locking through five high dams on the Cumberland, nine on the Tennessee. I love these rivers no less for their being sectioned off into lakes, and I've consumed the relatively cheap, clean electricity that their harnessed currents produce. Still, river rats and reservoir dogs alike should know what happens when rivers are dammed, what we lose as well as gain. Free-flowing rivers, those which run from source to mouth without a single dam, have been greatly reduced in the past one hundred years. According to the World Wide Fund for Nature (WWF), only 18 percent of North America's largest rivers (more than one thousand kilometers long) remain free flowing. A dammed river, though beneficial to us in short-term, quantifiable results—power generation, flood control, commercial navigation, water storage—brings about provisional, ecological, cultural, and spiritual loss. Water from a dammed reservoir is inferior as drinking water to that of a free-flowing river. Damming changes the habitat and alters what fish are harvested, sometimes wiping out entire species, particularly anadromous fish such as salmon or shad, who are blocked from spawning grounds upstream. Dams block sediment flow that is essential in sustaining wetlands and marshes. Most notably among tribal societies that are closely linked to waterways, dams alter the spiritual and cultural value of a river. When eight large dams were built on the Waikato River in New Zealand, the Maori tribe, who used the formerly free-flowing river as a cleansing, healing entity as well as a fishery, found that fish stocks on the dammed river were reduced. The elders of the tribe considered the river "ill" in its altered state, its strength diminished (WWF). While these spiritual and cultural effects are most evident among so-called primitive populations, losing free-flowing rivers affects us all, no matter how "civilized" or technologically advanced we are, regardless of whether we live in city, town, or countryside. What we lose spiritually, culturally, and aesthetically, we can't tabulate on a spreadsheet in dollars or kilowatts. Those of us born after our rivers were dammed don't even know what we lost.

While the era of frenzied dam building has probably come to a close, at least in the United States, it seemed worthwhile to me to articulate how a stilled river differs in character from its free-flowing, natural state. In paddling from reservoirs to transitional zones, I began to meditate on the components that make up the character of a river: history, geography, culture, and, yes, even people figured into this. Arriving at the point where a river revived itself, I thought, would enable me to see its rebirth and explore its personality differently than a trip from headwaters to mouth, the usual path. Along with that, I began to ponder the aesthetic issues of landscape. Why consider a river more beautiful than a lake? Why are some rivers judged more scenic than others? What's the relationship between aesthetic quality and the degree of human impact or manipulation of a place?

Four years I spent on this quest. I scoured the rivers and creeks of my childhood—Blood River and Clarks River in western Kentucky. In East Tennessee I ventured out in a pattern that resembled the spokes of a wheel centered on Maryville, where I live, to the Wolf River, the Big South Fork of the Cumberland, the Tellico, the Pigeon, and the Nantahala. In 2008 I got more organized and made a list of rivers from the east to west coast, a wacky driving/camping/kayaking loop de loop around America: the Gauley, the Connecticut, the Tippecanoe, the Columbia, the Bighorn, the King's River, the Dolores, the Brazos, and the James (in Missouri), a grandiose collection of names that resonated with me historically and aesthetically. In 2009 and 2010 I spent three days on the Edisto in South Carolina, went up Mayfield Creek off the Mississippi in western Kentucky, struggled against tiny Pistol Creek in Blount County, where I live, and returned to the Nantahala. Paddling upstream became a sort of religion, mostly solitary, with a vaporous deity that kept me searching. What did I find at these transitional zones besides the spot where the natural overtakes the contrived, the artificial? Depends. Some surprised me; some did not. Often, expecting much, I was disappointed. Expecting little, I staggered in wonderment at the perfect alignment of natural aesthetics. Some ended in a kind of victory, others in failure. And sometimes the failures, on reflection, revealed something of value, like an unearthed gem that needed polishing, a wine that needed aging. Other times, on these journeys, the zones, or the state of mind I hoped to attain by exploring them, appeared in unexpected places: on stagnant water, on a trail, in the car on the return home. All of these places, all of these trips represent, I hope, the possibility that the landscape has retained some mystery, that there exist within the scope of our daily travel places that can't be mapped and gridded and sold, experiences that can't be simulated in a video game. My first two books were narratives of trips with distinct beginnings and endings. This one, a collection of trips, includes starts and stops, escapes and confrontations, discoveries as varied and complex as the rivers themselves.

A SEASON BEREFT

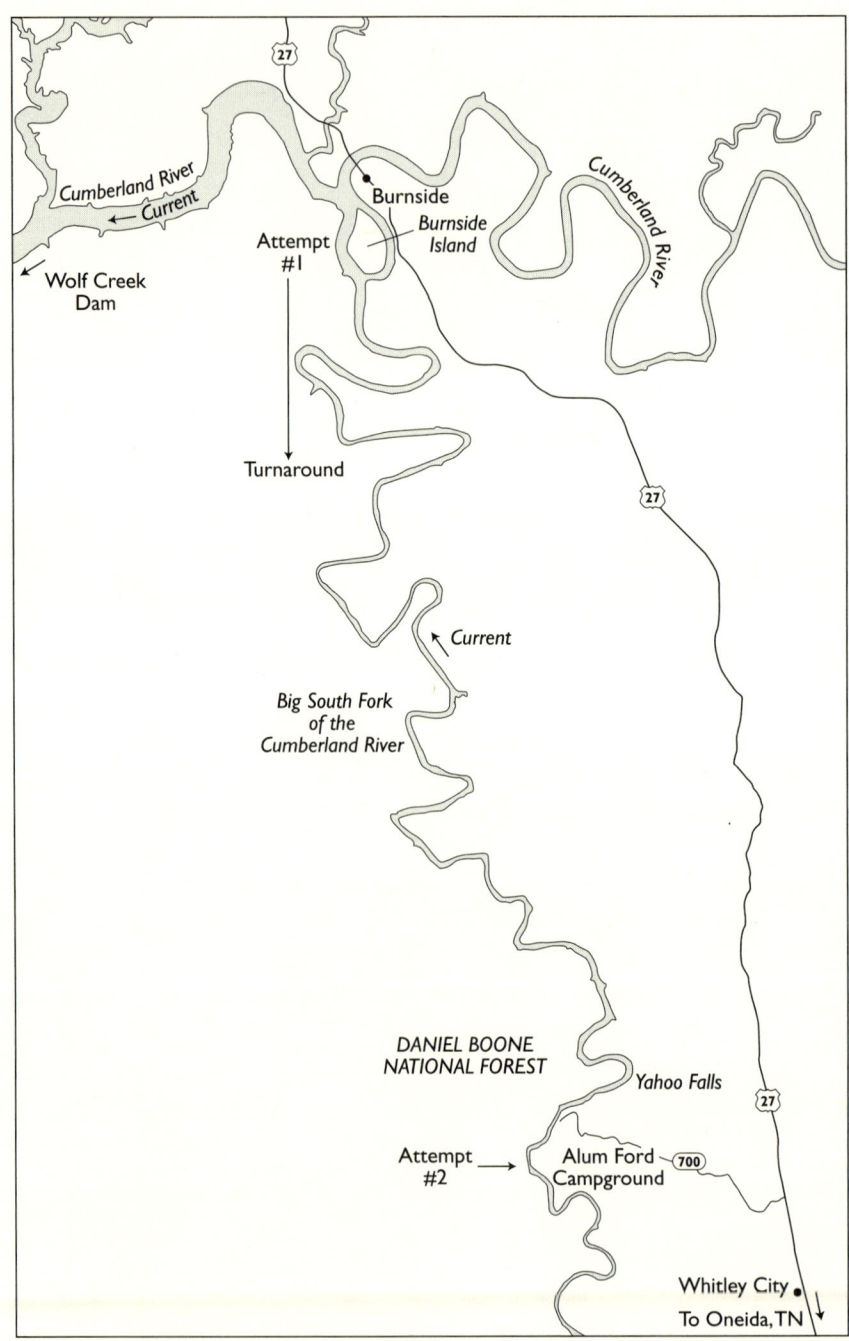

The Big South Fork of the Cumberland River

CHAPTER 1

THE BIG SOUTH FORK: PRODUCTIVE FAILURE

Drought, heat, and humidity plagued the summer of 2007, a good season to stay indoors and read about river trips rather than embark upon them. From late May onward, haze hung over the valleys and the rivers and obliterated the borders between forest and field, earth and sky, town and country, shoreline and watercourse, lakes and rivers. It was as if you were wearing some kind of netting or veil, particularly challenging for me, a guy seeking clarity, vividness, and distinct borders. The rivers and lakes of the Tennessee Valley began to shrink at an alarming rate; at one point, in July, the officials at the Jack Daniels distillery in south central Tennessee feared they would not be able to produce whisky that year, as their special spring had been reduced to a paltry trickle. Maryville's water supply dwindled to the point that the city asked for voluntary conservation and bought water from different parts of the county to avoid mandatory conservation. Gardens wilted; dust rose and hung in the air—the sun seared the grass brittle and brown. My dog had died. My uncle David, a survivor of the deadly Battle of Ia Drang, who did two tours of Vietnam, was slowly dying in an Oak Ridge hospital, where, between river trips, I visited him. He would joke around, no matter how dazed or in pain he was, and he'd always ask me if I didn't have anything better to do than to visit a hospital. I tried to explain what I was doing outside of visiting him, and he would usually say "uh huhh, uh huhh," with a bemused twinkle in his eyes, deadpan, and then he'd change the subject and ask me how my mom, his sister, was doing.

My routine: drive a couple of hours, set up camp in the evening, rise before dawn the next day and paddle upriver, away from the dam that formed the lake, until I reached a breaking point of mental or physical tolerance. Then I'd turn my bow, float back to the boat ramp, and drive home. The hellish atmosphere of that summer seemed to suit my unmoored temperament, and it did not bother me that no one else understood or really wanted to know much more about my watery excursions in a summer of drought. These two trips to the Big South Fork, a river whose mouth I'd passed on my journey down the length of the Cumberland, typify the elusive and mysterious nature of the liminal zones I sought.

Big South Fork of the Cumberland (BSF)

I'll just flat out say it: the Big South Fork of the Cumberland is haunted. That's my contention. I'm not a ghost hunter or one who uses psychic powers or reads cards of any kind, but I believe that what happens at a place stays with it, and anyone who has toured a battlefield knows this. I guess that makes just about every place haunted, though some resonate with past events more than others. Hauntedness is subjective, of course; certain people are more sensitive to it than others, perhaps more receptive. Sometimes, what you *think* happened at a place is as important as what actually happened there. It helps also to be alone, in the woods, and to stick it out all night somewhere remote from lights and traffic and electronic stimuli.

I headed to BSF twice that summer, the first time to Burnside, named after the Civil War general Ambrose Burnside, a blunderer as a commander but an innovator when it came to patterning his facial hair into a reversal of his surname: sideburns. Regardless of his failures, the most famous being at the horrible Battle of the Crater near Petersburg, Virginia, he looked, per author Shelby Foote, the part of a general, "a rather large man physically, about six feet tall, with a large face and a small head, and heavy side-whiskers." The town, named Point Isabel before Burnside's occupation of it, is in south central Kentucky, where the BSF meets the Cumberland River proper, though most have called it Lake Cumberland since the construction of Wolf Creek Dam, fifty-odd miles downriver. At the confluence of the two rivers, on Burnside Island, a lump of land not unlike the general's head, there's a state park. That summer, because Wolf Creek Dam was undergoing repairs, the lake had dropped below drought level, to the point that the boat ramp at the state park had to be extended for fishermen and pleasure boaters to launch their craft. I thought I could paddle from the confluence of the Cumberland and the BSF to the liminal zone in Tennessee, and I figured I could camp one night on the way upriver. Standing on the football-field-sized ramp and gesturing vaguely southward, I asked a car-bound ranger how far it was from here to where the "river started up again."

He glanced up at me and said, "Pretty far." Then he turned his attention to his radio.

I was not haunted at the state park campsite; I was harassed by the only species besides the mosquito for which I harbor unmitigated aversion. I don't want them exterminated, but I don't want them around me, either, and it seems that every time I camp and get ready to settle down, they show up and stay without an invitation. In a sense, I really am haunted by them. I am referring to the lowly, conniving, crafty, thieving *Procyon lotor,* more commonly known as the raccoon. Late that afternoon, as I sat at my picnic table, a lady in a straw cowboy hat strolled by and said, "Are you having trouble with raccoons?" I said no, I had just arrived.

"They're bad around here," she said. "Yesterday one came up and took a pop can right out of my hand."

That sounded a lot worse than anything I'd witnessed, and at the time I laughed inwardly and suspected she might be exaggerating for dramatic effect. Her use of the designation "pop" for soft drinks or "cokes," as southerners call them, exposed her as an alien to these parts. "I'll sure watch out for them," I said, and thought nothing more about the well-intentioned northerner's advice. Burnside himself was up to such advice during his stay here, long enough to build roads of cedar logs in order to fortify the banks of the Cumberland against Confederate incursions that never occurred. It was not until 1890, after the arrival of a railroad and increased commerce and population, that Point Isabel was renamed for the side-whiskered Union general. Poor Isabel.

After dark I lay in my hammock listening to the Cincinnati Reds play the St. Louis Cardinals, the Marty and Thom Brennaman father-son announcing team growing more and more disgusted with the Reds' play. Near the end of the game, when it was clear that the Cards would win and the announcers had resigned themselves to defeat, I dropped off into a fitful sleep, and my cheap radio spat static that woke me at intervals. Out of the corner of my eye, I saw something move about five feet away from me at the fire pit. I looked away and then back again and caught the raccoon peeking at me over the metal fire screen. I was too lazy to get up and try to scare him away, so I turned over and ignored him. About the time I dropped off, rustling and clanking awoke me. He was into my garbage bag, and this time I got up and approached him. The size of a small bear, he backed off, but not very quickly, as if he felt he had more right to the refuse than I did. I put the bag into the dumpster and all my food into the car, thinking that was the end of it. Later on, again on the verge of sleep, I saw movement at the picnic table, maybe fifteen feet away, and he was standing on his hind legs beside my gallon jug of water with the cap off. I ran him off again and poured out the water. Later that night, as the vandal darted from campsite to campsite, I slept little, wondering if he'd climb a tree above my hammock and jump me sometime that night. Coyotes howled in the distance, more civilized in their remote wildness than the marauding campground bully.

Strong headwinds and an underestimation of the distance to my next campsite, Alum Ford, thwarted my plans to discover the BSF liminal zone. I paddled ten hours that day, returning to the ramp sunburned (despite a gallon of sunscreen), dehydrated (despite a gallon of water), injured (I cut my ankle on the steep shale shoreline, nearly impossible to disembark upon at any point), stiff, and paddle weary. I did see a fox and her pups up on the steep rocky banks of the BSF, nor far from a house at the top of the bluff. She paused in her walk across the broken platters of shale and stared at me, reddish in the early morning light, her pups a lighter color, looking a bit like kittens from a distance. I didn't let the failure of my miscalculation prevent me from making another attempt at the BSF transitional zone.

Alum Ford and Yahoo Falls

By the next effort I'd thrown my back out after sitting on a rock for four hours in the Smoky Mountains. Julie and I had joined the pilgrimage to watch the synchronized fireflies that emerge for about a week each summer on a trail near Elkmont Campground. It was worth the trouble, the long wait for the emergence of the multitudes, their blinking choreographed, as if an expression of disbelief at so many well-behaved, reverent humans. Next morning, I could barely walk, and it was a week or so later, in early July, before I chanced a trip to Alum Ford, which had been my destination on the first BSF attempt.

For these trips I used state gazetteers, atlases that represent, in great detail, not only secondary roadways, but also rivers, lakes, and boat ramps. On the Kentucky gazetteer, it certainly looked like the lake narrowed to a river-like blue squiggle, but even gazetteers are not completely reliable, not for my errands. (And if they were, what fun would that be?) I hoped I wasn't overshooting the transitional zone, and I certainly didn't want to camp right beside it, but Alum Ford was the only campsite in the vicinity of where I thought I should put in. Hoping for peace from marauding raccoons in the deeper forest of the BSF recreation area, I had no idea what Alum Ford had in store for me.

I'd turned left off Highway 127 onto Highway 700 at Whitley City and began a winding course that not only led me downward in elevation but also seemingly backward in time. One house trailer, only a few feet from the road, had four or five cars parked in front of it, and Santa's sleigh and reindeer on the roof, the sleigh overturned and Santa nowhere in sight. Everyone had their brown yards mown to the quick, as if expecting a big rain to water their lawns beyond reclamation. The houses, sturdy and neat, were built with mixtures of brick and wood or Masonite. As I continued down, down toward the water, away from the houses, and entered Daniel Boone National Forest, I reached a sign that said "LAKE AHEAD," which in terms of my quest was a good thing. According to this sign, I could head upstream in quest of where the river began. I found the campground on a bluff above the ramp and chose a spot that overlooked the river, visible at intervals through the thick trunks of the trees. A chain-link fence, such that you'd see around a schoolyard or a prison, ran along the side of the slope a few yards below the campsite, as if to stop drunken campers from rolling all the way down the hill into the river.

One other site seemed to be occupied: a large, newish dome tent with lighter fluid, gasoline, and a gallon jug filled with water outside it. Hikers, I surmised, gone on a long trek. I walked down to the boat ramp. Sure enough, it looked like the winding lake of my previous excursion twenty miles downriver, except it was narrower and the bends were tighter. It had that same desolate look to it, the sharp and brittle tan rocks tumbling down thirty yards or so from the steep, forested bluffs, but here the bluffs were higher than downriver, and there were boulders the size of the Masonite houses I'd passed on the way, big jumbled rocks like the ones below Cumberland Falls, which you could hike to from here, if you had a couple of days, on Sheltowee Trace, the trail designated as such after the name the Shawnee gave Daniel Boone (Big Turtle). On its last ten feet, the ramp took a steeper grade and ended abruptly about three feet above the water line. Somebody had run over a harmless black snake, which lay crushed in the middle of the corrugated concrete slab.

After setting up camp, I hiked the two miles to Yahoo Falls while thunder rumbled all around. I knew nothing about Yahoo Falls, didn't even know it was nearby until I saw the trail and the sign. "Yahoo," to me, meant the sound you make on the way down from someplace high or something you said with your head sticking out the window of a moving vehicle; it was also a term that satirist Jonathan Swift coined for a species of isolated islanders who represented the baser, more primitive side of humanity, prone to violence, at the mercy of their passions, devoid of the capacity for reasoning. The campground signs advised to watch for bears and snakes. And also, they added, almost as an aside, don't hike barefoot and do take a flashlight if you go at night. I'd hiked a lot of trails and never seen that advice. Other signs, nearer camp, said you couldn't carry firearms into the area. My fellow Kentuckians had nearly obliterated the sign with bullets and etched a message with a Bowie knife, I'd guess, in wavering spindly letters: "F--k the Ranger." Three dead fish, whole, lay on the trail. They'd been there at least a week, keepers at one time. Would the doers of these deeds, my state brethren, be my camping companions through the long, dark night?

    I paused now and then to survey the cliffs looming over me, rivulets running down the rock, to my left the river, becalmed and empty. Then the disheartening sight of the river making a little roar that filled up the gorge. If I had stuck it out the other week, if I had kept going upriver another ten miles, that would be the first sign of current. Yet there was still the sign that said "LAKE AHEAD," and I was thinking I could paddle a few miles upriver before seeing another shoal like that one. I could pretend I hadn't seen this one.

    Nothing appeared to be falling from Yahoo Falls, though the rock shelter it had carved at the base was majestic. A sign told me that ancient peoples had dwelled here, and now no one was here but me. It looked like a good place for a rock concert—or a campsite for a hundred of your closest friends. A small family emerged antlike on a trail above, then disappeared, ascending, and a couple of guys peeked over the top, one hundred feet above me. I studied the space between the ledge at the top and the floor, and I finally found the water, a mist trickling over the edge and disappearing before it reached the solid rock.

    After the hike I walked down the boat ramp to the river, what there was of it, five feet at its deepest. Beside the ramp was a popular bank-fishing location, complete with drink containers, worm cartons, underwear, fishing line, etc. A layer of scooze floated on the river. I backed on into it and got my head wet, then climbed up the hill to my campsite.

    I waved hello as a kid drove past in a Volvo, and he waved back a bit skittishly, it seemed to me. He set up his tent at the end of the road. The neighbors with the gasoline and lighter fluid had not showed. A ranger pulled up in his SUV. I met him on the roadway, and he got out of his truck, a stocky young guy with a black pistol in a holster at his side. He asked if I'd seen my neighbors. The maintenance guys had told him the tent had been there for over a week, no one in sight. "Where could they be?" I wondered aloud. The ranger said the guys might have gotten arrested. Last year, he said, a group ran off into the woods away from the sheriff, who had a warrant for their arrest. They had the gall to file a complaint about some of their stuff being missing after they came back for it two weeks later. Shaking his head at this, he walked across the campground road and searched through the abandoned tent and around the vicinity of the site. Came out with what looked like a silver dumbbell. "What if he shows up?" I asked. "Do you want me to call you?"

"You can tell him to call me," he said. "I've got his lantern." This guy was from Arab (pronounced A-RAB), Alabama, and he'd been working in the Everglades when they transferred him here, closer to home. He said, "Good evenin'," and drove off.

With my macaroni and cheese dinner and a fire and my back not hurting so badly after the walk, I reached that calm nirvana that camping often produces. Frogs cranked up a chorus from the river and a whip-poor-will called from the bluff behind me. It was good to feed the fire and stare into it, no matter that I didn't need it for warmth. I liked the way it smelled, and the way you carried the smell with you the next day in your clothes and your hair. This was pine I was burning, a stack left by the previous occupants. It made a cheerful noise in the lonely woods.

When it started raining around eleven, the mysterious camper returned in an old SUV and put a rain flap over his tent. Then left again. I wondered if he noticed his lantern missing, and, if so, did he suspect me of taking it? He was a big shambling guy, but I couldn't tell much else about him in the rain and dark. I listened to the Reds lose to the A's, then to some kind of trucker product radio show, then later still to the Chicago news station. Even in the rain, sleepless, beset with existential darkness, I was glad to be out there. Dry, healthy, mostly at peace, I'd read Hemingway's story "Three Shots" earlier. Nick Adams, the main character, gets scared in the woods alone and fires three shots to summon his uncle and father, who are fishing on an island. He's just begun to understand the concept of mortality as a result of attending a funeral. He wonders when his "silver thread" will be broken. His father tells him there's nothing to be frightened of in the woods, not even lightning. But we know and Nick knows that what frightens you alone at night, in the elements, is primal and natural, that it goes to the core of our humanity, the curse and blessing of consciousness. Will the long-term camper return? The ranger? Are there restless spirits in this place, people who have died here violently or before their time? Does this place create malevolence or do Yahoos bring it here to enact it?

What's wrong with that kid down the way, camping by himself and driving that old Volvo? The most unlikely perps are the ones to watch. Stop thinking like that, I kept thinking. I turned up the radio and listened to truckers talk about fuel filters—desolate, disembodied voices keeping me company above a dying river. I missed the dead—my dog, my father, my uncle Ed, my aunt Emmy Lou, my grandparents, all of them—and I thought of summoning them and then feared that they'd come, as they have before, in my dreams. I recognized myself in them with a poignancy made sharper by my circumstances.

As I found out later, there was more than the mysterious camper and my gloomy temperament to account for my restless night. If I had parked in the lot at the top of Yahoo Falls, I would have read the plaque that commemorated the so-called massacre of 1810:

> Many Innocent Indian Women
> and Children who Knew No Wrong
> Were Massacred by Indian Fighters
> On August 10, 1810
> Let us Remember them
> With a Cherokee Tear
> In Loving Memory of Red Bird

As the story goes, the remnants of the Cherokee in this area had gathered at the rock house below the falls in preparation for an exodus down the Tellico Road to the Sequatchie Valley, where Reverend Gideon Blackburn, a Presbyterian, had a school on Cherokee land. The base of Yahoo Falls, at the big rock house, was the site of many speeches by tribal leaders, a well-known sacred place to the Cherokee. The daughter of recently murdered Chief Doublehead, Princess Cornblossom, and her husband, Big Jake, a half-Cherokee trader, had promised to rendezvous there and lead the members of the tribe south to sanctuary. Before the princess and Big Jake arrived, a local Indian fighter, Hiram Gregory, accompanied by a band of vigilantes, opened fire on the Indians, quickly killing the few braves and continuing to fire until the women and children lay dead or dying, one hundred Cherokee in all. Big Jake and Princess Cornblossom came upon this scene, positioned themselves advantageously, and picked off all but one of the vigilantes. The princess died of wounds a couple of days later, and Big Jake, it was said, died of grief-stricken madness.

This story was documented on several websites, one account written by an anthropologist, but it seemed just as many sites were adamant that the story was legend or hoax. There was no historical evidence of a massacre, asserted historians, white and native alike, and no one seemed to rebut the revelations exposing the story as fictional, though some of the characters, such as Doublehead and Jake Troxel, did exist. In fact, that fall Daniel Boone National Forest officials announced that the marker would be removed for two reasons: permission had not been granted by the forest service to erect the monument, and there was some question about "whether the event actually took place."

Regardless, something was awry there, not just in my head. There's something about a gorge that's dark and mysterious, especially a gorge in the South, where you don't expect it, where it seems to open up in land otherwise gently undulating, rounded off hills and hollows. I wondered how this deep cut of wilderness affected the people who live around it. Why would someone go to the trouble of making up such a story? Does the story contain a truth beyond whether it's factual or not? What does its existence say about those who imagined it and passed it down? What are we to think of those who accepted it? Of those who investigated its occurrence and found proof lacking? Not many came down here, the ranger had told me, and the campground was being threatened with closure. Some of the few who visited left the place trashed, ran over snakes, and shot up government signage. Did the gorge stir something in their souls, as it did mine? Does the thirsty river make them mourn the days it ran free, when Cherokee lived and hunted here?

The rain put out the fire. I must have slept some. At five that morning, two owls hooted my alarm, and I got up to kayak.

It did not feel good on my back to sit down in the boat, and the longer I was in the boat, the more my back expressed its displeasure. A beaver paddled toward me, then made a big splash as he dove underwater. A half mile upriver from the ramp, a shoal slowed me down, and I fought it a while, but a log lying across the channel narrowed the current and made it impossible to go forward. I drove out of the gorge, uncertain whether or not I'd return. The place repelled and attracted me, and I wanted to see the river with water in it, the tragic falls flowing over the ledge to the killing floor, where, as Princess Cornblossom was alleged to have said, while firing at Hiram and the others, "You kill our men. You kill our women and our babies. Their blood made red the land you steal." Or not.

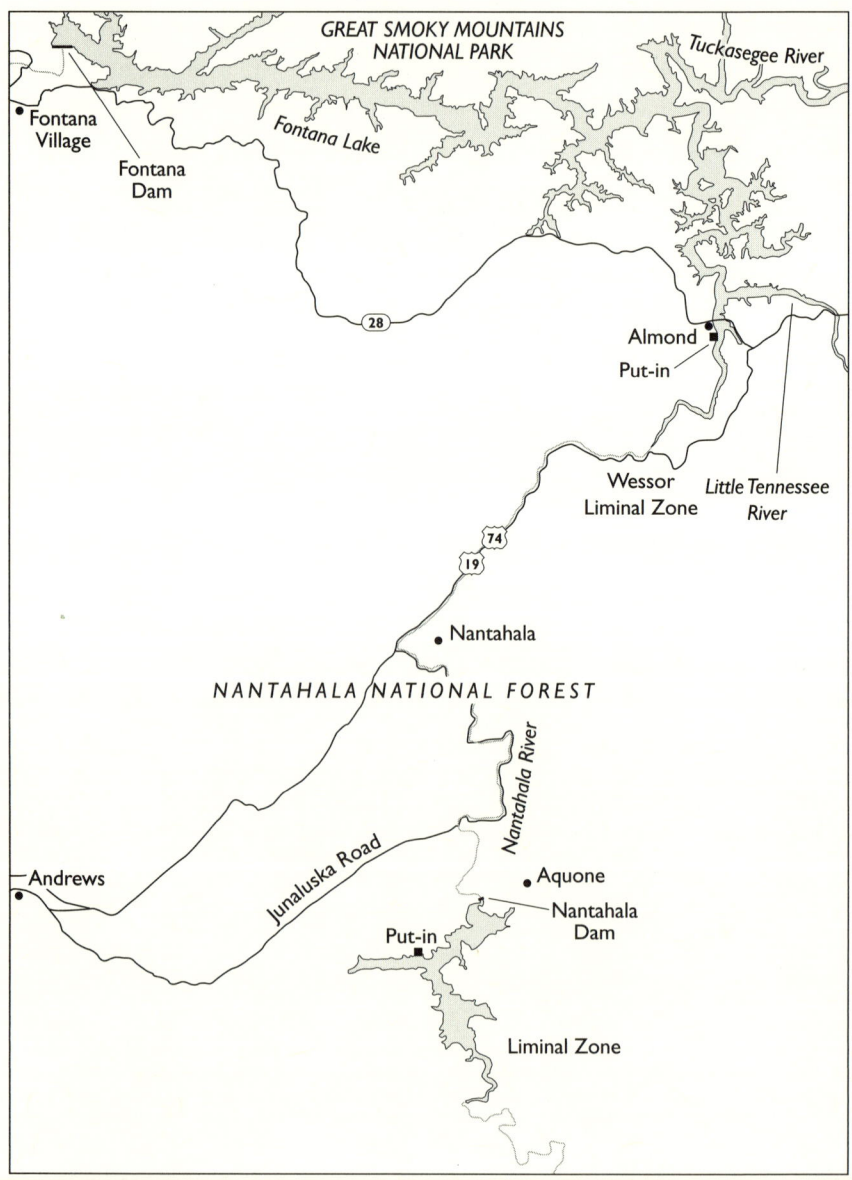

Nantahala and Fontana lakes

CHAPTER 2

THE NANTAHALA: THE LIMINAL UNVEILED

Nantahala I (Nantahala Lake)

The Nantahala River flowed through my memory like a ghost stream from another life. I'd taken a rafting trip down it thirty-some-odd years earlier with a church group my sister Melissa was heading up. I remembered following orders and paddling backward or forward when I was told. Our guide, a camp leader not much older than me and far from a professional river rat, recommended that we remain onboard at the falls, a six-foot drop at the end of the run near the Nantahala Outdoor Center (NOC), a gathering of shops and outfitters, with a footbridge for observers to ogle experts and blunderers alike. One of our crew did fall out, and we backpaddled to pick her up in an eddy below the falls. She was laughing, and I remember her laughter so vividly because I feared her dead or badly injured, the jolting of the falls was so powerful, the water such a violent maelstrom. In the summer of 2007, far removed from six-foot drops, church camps, and giddy Christian fellowship, I would put in at first on Nantahala Lake, above Nantahala Dam, the first dam on this popular whitewater river. A month later, in August, I'd put in below Wessor on Fontana Lake, formed by Fontana Dam.

After a two-hour drive over the Smokies alongside Lakes Chilhowee, Calderwood, and Fontana, I arrived at a hamlet called Aquone, where you could hardly get to Nantahala Lake it was so clogged with gated communities named Arrowhead Point and the like. At an ancient bait store that rented space for RVs only, a tough-looking teenage girl—angular, pierced, tattooed—turned to an old woman in a rocker and said, "Where could he camp in a tent?"

"In a *tent*?" asked the woman with a contempt that made me smile. She reminded me of my own grandmother, repeating my words in a gently mocking way to expose the absurdities of my plans. Then she told me a place I could camp, for free: Old River Road, actually Forest Service Road 308; "old river," it turns out, is what's left of the Nantahala River after most of it had been diverted into a pipe that runs parallel to the road. Here it's a little stream with rocks and islands barely big enough to stand on, topped with grass. I took the campsite with the least trash, next to the gravel road. Only a couple of cars passed in two hours. At this, the cleanest of the five or so sites, people had contributed their paper wipes and other refuse to the ambience. To the delight of swarming flies, somebody had poured their cooking grease on the ground. I covered that up with dead leaves and lit my citronella candles. The fire ring was full of ash and other refuse, including a slice of white bread that even the ants shunned. Above me loomed rhododendrons and maples and pines, high steep ridges on both sides of the river. Coming from the heat of the valley where I live, it seemed an oasis of cool freshness here, and it looked and smelled as if it had rained recently. That summer, even a rain that had fallen days earlier was worth relishing; stupidly, I wished for more.

I hiked toward the dam down the country road, partly retracing the steps of eighteenth-century botanist William Bartram, one of the first naturalists of European descent to document the flora and fauna of the New World, as well as the culture of the native peoples, including the Cherokee. Bartram traveled on foot twenty-four hundred miles across the Southeast. Not far from where I was walking, he came upon the Cherokee chief Attakullkulla (Little Carpenter), who had traveled to England himself in 1730, with the help of Sir Alexander Cuming. At the end of Old River Road, I walked past a recycling center on Junaluska Road, then began walking down Nantahala Dam Drive, which soon turned to gravel. After three miles I came to a floodgate, or emergency spillway, that was about twenty feet high and twenty to thirty yards wide. On the way back to camp, I could hear fat drops splatting on the leaves above me like shrapnel. At first I thought it was hail, as it was so loud. A guy in a truck came by and asked, "You need a ride or you just walking?" I told him I was getting some exercise. Big mistake. By the time I was at the highway, still a couple of miles from camp, a steady rain fell, then a sudden torrent that blurred out the landscape, a major weather event. I got beneath a big sycamore, but the drop-off of the ravine was too steep for me to get right next to the trunk.

After thirty minutes or so, the rain slacked off, and the swiftly moving clouds opened to reveal a patch of blue sky. The wind kicked up every now and then from different directions. That bit of blue in the sky was maybe the epicenter of the storm because the rain intensified and so did the thunder and lightning. Five or six jagged bolts struck the ridge right across from me, a quarter mile away. There was a North Carolina Department of Transportation compound across the road, all locked up, with a big fence around it, the shelters made of aluminum. My respect for nature grew as I stood there shivering and wishing the storm would slacken to the point I felt halfway comfortable walking through it. I was soaked and no longer amused by the storm, which had persisted for at least an hour. My teeth chattered as I took my shirt off and headed down the highway, striding fast to try to bring up my body temperature. At the recycling center an old guy eyed me, and I managed to say, "Good one, wasn't it?" He said, "Yeah, boy."

It rained lightly for the whole walk, not leveling off until I got back to the car, changed clothes, and cranked up the heater. I started a campfire with damp wood that spread smoke as thick as the fog that shrouded the ridges above me. The sky was shut out. The river had changed to chocolate brown and roared fast just a few feet from my campsite, a real river now, the diversion pipe notwithstanding.

Next morning, at 6:30, one other car was parked at the only public boat ramp I could find on the entire eastern shore. Above me rose high bluffs with majestic houses built of wood and stone and glass. There were no big boats at the slips, but quite a few kayaks and canoes. Up until now, I couldn't help it, I'd been thinking of the opening to *The Andy Griffith Show*, the virtuous sheriff and his son, Opie, walking down a dirt road with fishing gear while the bouncy, innocent theme song is whistled—no lyrics needed. That theme establishes the mood of Mayberry, North Carolina, code for the idyllic rural south, where poverty, crime, bigotry, and brutality are either nonexistent or nipped in the bud by the wise and gentle Sheriff Taylor. Now, having left the littered campsite, where the masses had recreated, I had gained access to another class, one with property, money, and leisure. Though remote from Mayberry, it was a beautiful place, I had to admit. Fog hung ragged above the water, and up in the hills it gathered along the receding and intersecting angles of the southern mountains. For the first two hours, there were no other boats, no human voices.

I had packed a change of clothes, tortillas, apples, life jacket, sunscreen, etc., but forgot the map of the lake, a fateful blunder. Gliding past a little island covered with fir trees, a hump of land rising above the water line,

I wondered what it looked like before the dam, whether it was too much of a hillock to bulldoze or if the dam builders decided to leave it for aesthetic relief. The dam was finished in 1941; it was, like other dams, justified by the war effort and the need for power. The town of Aquone lay somewhere below the placid surface, the familiar story of relocating families and farms and graves enacted here as in so many other places, history and landscape obliterated. Before that, the Cherokee were relocated from the same area, not by official flooding but by law, specifically the Indian Removal Act of 1830. In what would be known as Aquone, a stockade was built to hold the Cherokee that General Winfield Scott was able to capture and herd down from the mountains, and from there they were marched west to Oklahoma on what became known as the Trail of Tears.

I passed a cove on my left and kept on straight toward the end of what looked like the main lake. There was a compound around it of houses built on flat land with treeless lawns, not like the majestic houses built into the bluffs, these plainer and more utilitarian middle-class ranchers. At this point the lake widened, and the fog thickened to a degree where it seemed I was heading into some kind of oblivion. I had a moment of profound sadness, a feeling of inexplicable loss—but I paddled on, and gradually the bluffs emerged again, and the feeling dropped away. I checked out a couple of coves at the end of the lake, but both were dead ends; I felt like a blind man groping without the map. One cove ended at a shack with a rebel flag and a big NO TRESPASSING sign. I paddled back down the lake, disheartened, realizing at this point that I should have taken the first sharp left instead of going straight, that if I'd had the map I would have realized this. Two hours of paddling wasted. Somebody was arguing with her kid at the car. Really yelling at him. And I had been thinking how perfect it would be to live on a lake so cool and clear and quiet, but no matter where you live, there's still trouble. Humans create it among themselves, families, neighbors, whatever. We seem to crave it. Then I saw a couple of kids kneeling over something at a boat ramp, examining it, not talking, so quiet. I wondered what fascinated them so. Across the way somebody started hammering. A couple of other voices carried across the water, and car doors slammed, people going to work to attempt to pay for these big houses.

Paddling back toward the ramp for two miles, I reached the cove I should have taken, the lake bending hard right. Father up the cove, the water turned turquoise, different from the slate gray of the main lake. Dead limbs of fallen trees reached out from the banks, distinct in depths of several yards.

As the lake arm narrowed to thirty yards or so, an amber film lay across the surface. Aquatic weeds appeared on one bank, bordering a field of bushes that seemed to thrive while rooted below the shallow water. I thought I saw the scooze stir in some current but couldn't be sure. The water had turned from transparent turquoise to the color of strong tea. Somebody had a piece of land for sale, flat and full of pines, a johnboat pulled up and tied to a tree. The bottom appeared, about five feet below, and the passage narrowed to maybe twenty yards across. The new coolness of the river soothed my calves through the plastic hull. There was a house with some steps leading down to a platform ten feet above the water, two padded chairs on the platform. Here these folks could get up and see where the river begins. No boats, no inner tubes, nothing here but the chairs. Up ahead the real river charged toward me across some shoals so shallow I had to raise my rudder. Farther up, the Nantahala roared low like static. I paddled hard and fast until I scraped rocks and drifted back downstream.

Driving on a back road toward Andrews, I turned on an AM radio broadcast, and an old fellow, Bill, was asking his listeners the best way to can beans. One caller repeated what her mother was shouting from the background, and I imagined her in a kitchen, on a stool, wiry and wrinkled and stern, something fragrant steaming on the stove.

Bill said, "I bet she doesn't have to can anymore."

"No, but she sure knows how to tell me to do it," said the daughter.

Some guy called who had known Bill's mother. He remembered how tough she was, that she could grow anything. "She used to sit out on that porch in the sun no matter how hot it was," he said. He had a high girly laugh that Bill let us listen to.

"She thought the world of you," Bill said to the caller.

A woman called, asking Bill if he'd bought some ice at the grocery store, that she thought she'd seen him there earlier. He said that he did indeed buy ice at the grocery store but didn't see her. She'd gotten her hair cut short, she said, and Bill exclaimed, "I would have spoke if I'd known you."

As I drove past the grandiose entrance to Fields of the Wood Bible Park, near Ducktown, static broke over Bill's soft voice. He was trying to give away some potatoes he'd grown, and he was asking if anyone had any homegrown tomatoes for trade. I was not far from Hiwassee Dam, where I had taken Jasper on his last long road trip. I'd put the backseat down for him so he would have plenty of room, but he had trouble standing up and swaying with the curvy roads, as he had done with such ease before cancer had enfeebled him. At the tail waters below the dam, he poked around a bit, but he was too hot to explore much, and he was not strong enough to swim. I didn't have the heart to revisit the dam and watch the kind of crazy reincarnation of current you see in tail waters. There at the borderland between North Carolina and Tennessee, where some country singer long gone usurped the airwaves of the companionable radio host, remembering Jasper was enough. I drove over the Unicoi Mountains and headed toward home.

Nantahala II: Fontana Lake

A month later, in August, on the hottest day of a hot summer, the heat index over one hundred, I set up camp at Tsali, where dozens of miles of mountain biking trails wound through the woods above Fontana Lake. Next day I arose before dawn, not having slept well, and drove the two miles to Lemons Point boat ramp, big enough to launch a naval destroyer, though I was the lone utilizer on this morning. There are reasons I like to launch at such an uncivilized hour: to maximize the distance I can cover in a day; to avoid the heat as long as possible; to have the lake/river to myself; and, having the river to myself, to enjoy the fringe benefits, such as startling fish and other wildlife who don't expect humans ambulating at such an hour. I paddled past a marina, hunkered low and still in the early morning haze, nobody stirring, no motors idling with blue smoke, no rods banging against plastic or aluminum. A fisherman cast his lure in the shadows of a railroad bridge, the line a whizzing thread above the golden-lit vapors that rose from the water. A mile farther, some kind of machine, invisible to me, devoured limbs and greenery along the border of what I assumed was a road or a railway. That was the extent of the activity: fish didn't jump, and nothing stirred in the forest. By this time I was tiring and getting cranky; the lake was narrowing and starting to bend at an extreme angle in the distance, but the water was warm and untroubled, and I saw nothing that resembled a river. I thought about turning back and napping in camp before it got too hot.

The night before, I had been surrounded by conversers. One of a pair of sleek biker guys was talking on his cell phone to a recently divorced friend named Wiffleberg, who was ready to party, who wanted them to drive the half hour to Wesser, at the Nantahala Outdoor Center, where there was a contest among the river guides. Later, one of the mountain biker guys asked me if I was getting an Internet signal. I was not. A husband with a wild Edward Abbey beard, his wife, and adolescent son occupied an RV across the way, the father and son serious kayakers, it seemed, who had a sarcastic rapport, the wife, constantly cooking or cleaning, mostly silent. I could see their legs from the calf down under the RV, father and son lounging in chairs in their river sandals, Mom moving back and forth in jogging shoes, always on her feet. I'd driven to NOC after setting up camp and refreshed my memory about Lesser Wessor and Greater Wessor Falls, the big parking lot, the restaurants, the outfitter stores with boats and paddles and endless gear. A footbridge straddled the river, and flat-bellied young men with eyes for the water muscled kayaks smaller than army cots up and back in the angry water; strong and confident young women, effortlessly beautiful, some just as good, if not better than the men as river guides, negotiated with the powerful currents that flowed through Wessor, some at the stern of rubber rafts, others flitting about in the crashing water in shoelike kayaks.

  I hiked away from the hubbub down the river to Greater Wessor Falls, which looked unrunnable, a tumble of rocks and frothing current with no clear passage. The Greater Wessor is runnable, though considered by some a class V, which means you better know what you're doing if you want to brag about it later. I hopped onto a railroad track, creosote fumes rising from the blackened crossties simmering in the heat. Below, in swampy clearings next to the river, stood pods of tents, a community of sorts, where wet suits were hung out to dry, a campfire or two smoldered, beer cans and bottles scattered here and there. Beyond this I hiked, the river beside me. It had calmed considerably since Greater Wessor's complex fury but still flowed and flashed its teeth against deadfall. I hiked until the river curved right and the railroad veered left. At the shoreline, a concrete wall—perhaps once a bridge—extended partway from the near and far bank. Willows grew from the shallows. Current swirled around the gray trunks of dead trees midriver. Beyond, upriver, it looked like a lake, acted like a lake. I turned away quickly and hiked sweatily down the track back to NOC, afraid that I'd accidentally discovered what I'd driven all this way to find in the boat. Surely not, I thought, and I forgot about it, trying to sleep at the hipster campground among the late-sleeping mountain bikers, who operated their cell phones, halogen headlights, and laptops deep into the night.

  An hour after dawn, I was beyond boredom and crankiness. I had entered a semitranscendent zone with which I was familiar on long trips, where nothing existed but the movement of my paddle, my eyes unfocused and glazed.

Hints of change snapped my vision into focus. Bubbles popped to the surface like signals, all across the lake, and it looked like the patter of raindrops, though precipitation had been so rare, I held my palm up to confirm the heavens' continued continence. From the bubbling water arose a fishy odor. A long grassy point appeared at the next left bend, two skeletal gray trees standing out on it alone. Gelatin-like tumors grew on dead limbs and on rocks. The bottom came up. I saw cans. Shoes. Rocks. I imagined screams, squeals of delight, rafters tossed by a river I was approaching. I could read the labels on the cans ten feet below me.

Sometimes you see things in the outdoors that make you blink or rub your eyes, things that seem incongruous, contrary to the rules of normal existence. At times phenomena like this arise on rivers. Camped beside the Cumberland River, photographer Randy Russell and I peered out our tents after a four-hour deluge, and it looked like Christmas lights had fallen on the river; fireflies had been struck down by the storm, dazed and blinking as the floodwaters carried them toward Cumberland Falls. On the Clinch River, Uncle David and I saw a buck plunge into the water and swim right across our bow; we followed, keeping pace, close enough to see its breath in vaporous puffs until David stopped paddling to spare the deer its panic and let it cross at its own pace. It was uncanny to be able to keep up with something you normally think of as fleeting, impossible to catch, and it took ten minutes for us to realize we might be causing him harm. That same kind of anomaly emerged on the Nantahala. Here it was a complete change not only in the lake and the landscape around it, but in the atmosphere itself. As I finished rounding the bend, a solid fog bank about three feet high emerged across the width of the river, its border so distinct that for a moment I wondered whether I should enter, and if I did, where I would emerge from it. I quieted my paddle strokes as the fog enveloped me. The air was refrigerated, 20 degrees cooler than the smothering heat behind me. I recognized the concrete structures of the previous day's hike, seemingly farther apart in the obscurity of the vapor, and instead of being disappointed that I'd already been there on land, I was surprised that I'd been able to reach this place, which seemed so remote from the landing-strip-sized boat ramp and marina that I'd started out from that morning. I paddled fifty yards past the structures against the current, the river deep and strong here, and came to an island that split the river into forceful twins that turned me back.

As I let the river carry me back, the sun strengthened enough to pierce the fog, striking the surface at an angle I hadn't seen going upstream. It was if a curtain had been drawn. Carp, dozens of them: these were the bubble makers. They swam right under the boat, fearless, a foot to a foot and a half long. Their square scales sparkled in the sunlight, and these lowly fish for which I'd had nothing but contempt and disgust now looked magical, and the liminal zone lent them dignity and an organic sense of belonging that I never would have considered. A good-sized turtle poked his head out of the water, looked at me, and said, "Psssshhhhhhht."

On the way back, two kayakers approached, lean women in their sixties.

"Are y'all from around here?" I asked.

"Yes," one said curtly. "Are you?"

I said no, and they smiled at each other. I asked them about the concrete structure. One with a floppy hat said it was an old railroad bridge.

"There were whole communities here," she said, "before the dams."

Fontana Dam, downstream from us, formed the lake we paddled. It was finished in 1944, built to generate power for the war effort. The towns the woman referred to were Fontana, Bushnell, Forney, and Judson; more than thirteen hundred families had to move. Upstream from us, Nantahala Dam, finished in 1941, released water from its powerhouse at a level to maintain the rapids that fueled the recreational rafting, canoeing, and kayaking businesses in Wessor and elsewhere.

Instead of commiserating about the dams, I marveled aloud to the women about the fog, the carp, and the turtles up ahead at the liminal zone, though I didn't use the term *liminal*. They nodded and paddled on, either disbelieving me or unmoved by my description. I restrained myself from paddling after them and trying to explain about the wisecracking turtle, the refrigerated air, the glamorous carp, the magic of the transition. Perhaps they already knew. Ironic that a dam that destroyed so much made this place, this moment possible.

Aside from exploring the BSF and the Nanty that summer, I went up the Wolf, the Blood River, and the Tellico River that summer. The Tellico was the closest to home, in the next county, and in the next chapter I describe that first trip within a more extensive meditation upon dams.

CHAPTER 3

MY HISTORY WITH DAMS

Among a class of bright students, you can have a pretty good debate about dams—their benefits, the damage they do—and it helps if you transport students somewhere that they can actually examine one of these engineering wonders. What I do during the first week of my Maryville College naturewriting class, "Words and the Land," is break the class into pairs. One person in each pair has to put on a blindfold in the college van ten miles away from Fort Loudoun Dam, and then the other leads the blindfolded charge down the sidewalk that approaches the powerhouse on the lower side of the dam. The idea is for the blindfolded students to try to use senses other than sight to guess where they are. We go in January, but even then the blindfolded usually smell something fishy, an aroma that is strong below dams, where the turbulence of water coming through the penstocks stirs up food sources for a variety of species. Certainly they hear the sound of water lapping against rocks, the cries of gulls, the squawk of a heron, maybe a fisherman grumbling about the line of giggling, stumbling young people disturbing his quest to snag a giant catfish. A couple of times the siren has sounded to warn fishermen below the dam that someone is locking through and that water will be released from the lock, creating turbulence.

Usually one of the blindfolded students guesses correctly after a few moments of contemplation. Sometimes it's a local who had an idea where we were going in the first place, sometimes a kid who has a sharp sense of geography and space and water. I've often wondered if students can "feel" the dam looming above them, the sense of artificial space created by a large structure bringing a river to a halt and containing a world of water.

The fishing pier below Fort Loudon Dam affords a good view of the works: the churning water below the powerhouse, the riprap reinforcing the bank, the scale of the structure, its sheer strength and mass. There's even a rusty turbine on display in the parking lot, and at the top of the hill a power station that shows the students, without me having to explain much, that the dam uses water to generate electricity and, in generating power, it's not emitting any visible pollution such as smoke. (Visible pollution lies all around us, unfortunately, a testament to the popularity of the place and the environmental ethics of some who visit; sometimes a conscientious student will grab some trash—a wad of fishing line, a plastic bottle—and put it into a nearby trash can.) Though we're on the opposite side of the dam from the lock, sometimes we get lucky, and we're able to see a barge locking through going downstream, away from Knoxville toward Chattanooga or beyond. Students can see that although the dam forms a barrier to navigating vessels (and fish), it also accommodates passage, and its pooling creates a channel of consistent depth that can be controlled by releases from the dam. We can see the floodgates, almost never open, and talk about the dropping of the reservoirs to winter levels in preparation for winter and early spring flooding. On the drive back over the dam, they can see the marina above the dam, on the reservoir, and they can look down into the lock chamber where, I tell them, you feel like you're sitting in a draining bathtub as it lowers you from one level to the next.

Then we take the three-mile drive over to Tellico Dam to complicate our discussion further. Here, usually without blindfolds, we just walk out on the top of the earthen part of the dam toward the structure, a stroll of a half mile or so. On one side of the sidewalk, bordering broad Tellico Lake is a swimming area, deserted and dry in January, at winter levels. On the other side, if we're lucky, we might see a deer in the open field and beyond that a highway and the Tennessee River a few hundred yards below where we had been standing at Fort Loudoun Dam. We pause together next to the locked chain-link gate at Tellico Dam proper, where a sign tells us there is no trespassing, that this is the property of the U.S. government. At the top of the fence are coils of barbed wire. First I ask them what river was dammed here, and just about always someone says the Tellico River; rarely do students, even locals, realize that it's the Little Tennessee River that became Tellico Lake, that the Tellico River is an upstream tributary of the Little T. I ask them what's different about this dam from the other one and what's the same. The flood gates are a similarity, as are the banks reinforced with riprap, the helmet-sized limestone chunks, imported by the tons. Usually they notice first of all that there's no lock for navigation, no power generation here. They see the difference in what was created above the dam and what's left of the river below, where fishermen are always gathered, no matter the weather. "Why is it here?" I ask them. "Why was this dam built?" They are hard pressed for answers, as am I.

Often one student, gesturing at the lake behind the dam, will say the dam made the big gated housing communities possible, and though none is visible from this vantage point, we have passed one of the grand entrances of stone and wrought iron—Rarity Pointe—on the drive to the dams. I tell them that high-end development certainly seems to have sprung up above the pooling of the Tellico and Little Tennessee Rivers, but that originally the plan was for economic development of a different sort, the reservoir intended as a draw for industries and jobs. As Jack Neely pointed out in a 2004 *Metro Pulse* article on the twenty-fifth anniversary of Tellico Dam, the economic development hadn't come to pass, at least to the extent it was promised. At one time a futuristic city, Timberlake, was planned for the reservoir. It was supposed to create twenty-five thousand jobs, TVA claimed, in twenty-five years. Twenty five years later, Neely pointed out, the actual economic gains were more modest: four thousand jobs in Monroe County. "That's part of what we gained," I say to students. "What do you think was lost?"

If there's a paddler in the class, he or she will answer that a free-flowing stream was lost, somewhere interesting to kayak or canoe, and I might ask the student what's wrong with paddling on a lake and also point out that if you travel far enough up the lake that you'll find that the Tellico revives itself into one of the best whitewater runs in the state. That said, you've got to drive a lot farther to see the multidammed Little T come back to life, not far from my search for the Nantahala's transitional zone in North Carolina. Somebody else might wonder what happened to the land that lay underwater, who owned it before the dams, and how much was paid for it. A student, often a history major, might mentioned the burial sites of the Cherokee that were flooded by the dam, and the fight over a small fish known as the snail darter, which environmentalists claimed would be rendered extinct by the transformation of the rivers into a lake. Somebody might also know that the Tellico was prime trout fishing before the dam was built, that people farmed along its banks, and that some are still not happy about what happened to their land and what they were paid for it. Usually we end up wondering why the dam was built in the first place, and we contemplate the place without it. We rely on our imaginations to envision the landscape without the dams. It's not easy. Students' line of sight is typically drawn to the space below the dam, where a version of the river reappears for about a hundred yards before it meets the Tennessee River. Narrow enough to throw a stone across, it looks like a river, but it doesn't exactly move like one.

I've been teaching this course for seven years as of this writing, and I never tire of visiting the dams, not just for the good discussions and being able to get outside the classroom, but because dams fascinate me. I'd rather they weren't there, I mourn the fact that they kill long stretches of great rivers, that they bury the spirit of a river, but I also can't help marveling at the human effort, the engineering feat that it takes to create even an average-sized dam such as Fort Loudoun, which, if you lock through, drops you seventy feet from the lake above to the lake below, called Watts Bar. I'm drawn also because of the mystery that dams create, the sense of danger, the myths about catfish as big as seventies-era sedans, the swirling whirlpools below the powerhouse, and the tantalizing and terrible possibility of failure and flood, a fairly frequent occurrence worldwide and in the United States, most famously in the 1889 Johnstown Flood, when the South Fork Dam failed and twenty-two hundred died. I've locked through fourteen dams in all, on the Cumberland and Tennessee, each of them in my canoe. I locked through Pickwick Dam at night with Jasper, not a pleasant experience, especially since we had to paddle from the lock across the roiling tail waters, Jasper whining and fidgeting in the bow, to the place where we needed to camp. Wilson Dam, near Florence, Alabama, was the most bizarre, most beautiful, most disconcerting dam I have approached. Unlike most dams with the big doors that open up to the lock chamber, Wilson's chamber is sealed with what looked to me like an oversized cattle gate that stood about fifteen feet above the water line. If the lockmaster hadn't come out and asked me if I wanted to lock through, I might have backpaddled on out of there. In Wilson's lock we were lowered ninety-four feet, the longest drop of any dam east of the Rockies. Aside from the Pickwick experience, which was really no one's fault but my own, locking through dams has been a nice break on journeys, interesting, if not a little intimidating, with the lock operators always affable and professional.

It's hard to measure the aesthetic damage a dam does with any kind of objective list of grievances, though Patrick McCully in his book *Silenced Rivers* does a good job of establishing some criteria:

> Nothing alters a river as totally as a dam. A reservoir is the antithesis of a river—the essence of a river is that it flows, the essence of a reservoir that it is still. A wild river is dynamic, forever changing—eroding its bed, depositing silt, seeking a new course, bursting its banks, drying up. A dam is monumentally static; it tries to bring a river under control, to regulate its seasonal pattern of floods and low flows. A dam traps sediments and nutriments, alters the river's temperature and chemistry, and upsets the geological processes of erosion and deposition through which the river sculpts the surrounding land. (10)

That's a fairly comprehensive summary of the changes, though it's geared more toward ecological effects than aesthetic judgment. To make such a judgment with authority, it would help to know how a particular river looked before it was dammed, what was lost visually with the flooding of a valley, the stilling of waters. Since most dams around here were built in the 1940s or earlier, it's becoming more and more difficult to find anyone who remembers life near the water before a given dam, somebody who could talk about when a river ran where a reservoir now stands.

In 1998 I found one guy on the Tennessee River who still harbored bitterness about Watts Bar Dam. After Jasper and I had canoed almost a week down the Tennessee, we stopped for lunch at the shoreline of John McMurry's farm, situated at the confluence of the Tennessee and the Clinch Rivers, with a beautiful two-story house built in the 1880s, red barns, cattle that grazed pasture that sloped gently up from the lake. I asked whether he acknowledged, among other benefits, that Watts Bar and other dams had prevented floods. He looked out over the expanse of water that had once been pastureland, extended his arm, and said, "What do you call that?" It wasn't an explicit aesthetic judgment, but the implications were clear, his criticism based on appearance. A flood can leave a lot of water standing where people and animals once occupied land; it covers up components of the landscape that were not only economically valuable but that were a pleasure to behold—cows, grass, limestone outcroppings, trees, sloping terrain. I could tell students about McMurry and show them grainy black-and-white photos of the Tennessee River before Fort Loudoun Dam, but the dam in the area that was built most recently was

Tellico—finished in 1979—and that was the first place I started in my quest for transitional zones and the one place where people could still describe what the river looked like before the dam was built.

Francis Brown Dorward, who fought against the building of Tellico Dam, has gathered testimonies in her book *Dam Greed* from people who lived in Monroe, Blount, and Loudon Counties at the time the Tennessee Valley Authority began buying up land in preparation for the flooding of fourteen thousand acres, pooling thirty-three miles of what was left of the free-flowing Little Tennessee River. To me the book is valuable because it brings together so many recollections of the Little T before it was dammed, something that exists only in the memories of those who lived along its banks. John Lackey Jr. remembered floating down the river, catching native brown trout and rainbow trout that the Tennessee Game and Fish Commission had stocked in water where the temperature was a chilling forty-five to sixty-five degrees. He would fry the fish at the end of a float at the Rose Island picnic area (18). The mayor of Vonore, Fred "Fizz" Tallent, said he "had a favorite fishing hole under a shade tree where I would sleep.... In the spring when the fish were running, I could catch dozen [*sic*] of varieties. That clear, clean, cold water made delicious fish" (79). The river, Ben Snyder said, "stirred my imagination. I could picture Indians making the fish traps, standing on the rock traps, and spearing or netting the fish for drying or eating" (105). Shirley McCollum Brown remembered watching the river pool and grow still after the gates of Tellico Dam were closed: "Each day, I watched the slow death of the once free-flowing river. I was so sad the tears welled up and my throat hurt" (107). Junior Pugh said he wanted his grandchildren to "remember . . . the rich sandy soil that was as fine as the Nile Valley of Egypt before it was flooded" (113). Mayor Tallent eloquently described the loss of knowledge that comes with the obliteration of a river: "The people at Rarity Bay don't even know what river was destroyed to make the lake. Their fancy seven-color map calls the dam the Tellico River Dam. It was the Little Tennessee River that was dammed. The Tellico River flowed into the Little Tennessee River and was not backed up much" (81). Muriel Shadow Mayfield summed up the bitterness that many feel over TVA's exercise of eminent domain and subsequent flooding of rich bottomland: "Our fight was lost. But I hope people will someday value pure, clean, free-flowing water and rich, deep, fertile soil" (102).

Not everyone is angry about Tellico Dam. One writer, Jim Thompson, directly addressed the aesthetic superiority of the artificial reservoir over the river: "The lake is blue, still and beautiful. Above the mist, a line of blue mountains guard the southeastern shore and in the sky above, the storm clouds have dissipated over Tellico Lake. The future beckons" (*Tellico Dam and the Snail Darter,* 1991).

The farther up the lake you go from the gated communities, the golf courses, and the dam, the more difficult it is to find a boat ramp or a road where you can put in a kayak. It's easy to find a boat ramp on the part of the lake that Jim Thompson was talking about, where the water is deeper and big marinas are tucked into the big coves that were once creeks. It's also easy to access the Tellico River where it rages out of the mountains. Bob Lantz in his book *Tennessee Rivers* stated that when the water is at the right level, the Tellico is one of the premiere whitewater runs in the state: swift, clean, and technical. In my quest for the Tellico River's transitional zone, I was looking for a place in between these two personalities, a place that most people don't know or care about, and my tool was the Tennessee gazetteer, with its detailed rendering of county back roads.

Late that spring, after Jasper died, I spent two weeks driving around Monroe County with the gazetteer open in the front seat beside me, the yellow kayak lashed to the roof rack. Most of the time, I puttered around on Ballplay Road and Belltown Road. After three kayak trips that ended in failure as a result of having put in too far up the lake to reach the liminal zone, I spotted a road off Belltown that seemed a likely candidate. From the map the little road looked like it ended at the river. Narrow, patchy pavement led me past a passel of house trailers then dove on a hairpin down a rutted dirt road and ended up at a grassy circle where locals had been fishing off the bank. At 8:30 on this cool May morning, the fog had just lifted, and it was so quiet that the bumblebees seemed to roar. Woodpeckers ratatatted and called out shrill and roosterlike, one pecking at hammer-and-nail speed, the other going full-tilt staccato. This was the kind of place where I worried a bit about leaving my car. It was not posted as private property, but it clearly was not what one would term public access. I set the kayak down in the water next to a discarded container of night crawlers, a place that bank fishermen had frequented to the point that it leveled off close to the surface of the water. I could hang onto the bank as I slid ever so carefully into the cockpit of the boat. I rustled the thicket of weeds with my stern as I settled in and paddled a stroke or two out from the bank, letting the little boat glide through waters that reflected the cloudless sky. Here the Tellico had the look of a wetlands, the banks seventy-five to one hundred yards across, stumps and dead trees protruding from the surface or just a few inches under the water, cattails thriving near the banks, thick forest on each side of the river. Only by squinting could I discern the stumps and limbs that plotted my demise just below the surface, dark shapes like shadows. Sycamores, walnuts, maples, and oaks leaned out far enough for me to paddle in the shade. It looked like a river, but there was no current, absolutely no movement, on this calm day.

 For the first couple of miles there were no houses or boat docks, then scant evidence of humanity: a rope swing, a primitive wooden dock, a lawn chair or two, a forked stick for a pole. A deer snorted from the bank, and, instead of taking off, he stayed there, somewhere in the dark woods, snorting more and more shrilly. I paddled through the intimacy of the morning feeling like a clumsy intruder. Cane Creek on my left was clogged with deadfall. After a few miles the river/reservoir began to narrow so much that it resembled what I would call a creek. I went through a couple of places where trees had fallen across the entire river channel, and there was a narrow passage I had to duck under. White clusters of mountain laurel bloomed along the bank. The woods were thick and dark, and alongside the cattails stood dense stands of cane. Still, there was no sign of movement in the water.

 After so much remoteness, the concrete ramp surprised me, especially after I'd spent so much time driving around looking for access. This was one boat ramp (Big Creek Ramp) not marked on my map. I got out and walked up the broad gravel parking lot to the road. Goats milled in a small pasture across a creek. No road sign. No sign of cars. One house on a hill a mile away.

 By midmorning the trees reached across and met above me, the river constricted to thirty or forty feet. I looked for signs of current in floating leaves or tree limbs that dipped onto the surface of the water. Against a fallen log that tilted toward me, there was a stirring. In the bend a pile of deadfall had trapped a child's plastic four-wheeler, the first substantial trash I'd seen, other than the usual fishing line and cans of corn (trout bait) at the boat ramp. Another mile and the river bottom came up to meet me, about four or five feet below, and I could see it clearly as I moved against the current, not strong enough to halt my easy progress upstream. After the shallows it became deep again, and the current was imperceptible. There was farmland on both sides, mostly pasture, and I smelled the freshly mown hay. Every now and then, I'd hear a tractor or a generator burp to life. Farther on, I began to hear cars on a highway. No trumpets blowing, no fanfare, no solid bank of fog, as on the Nanty, but I'd found where another river revived itself, farmland on both sides, and I felt as if I knew something intimate about the Tellico.

This place I'd discovered seemed so remote from the dam where I took students, where the lake was its broadest. It also differed from the confluence of the Tellico and the Little Tennessee, twenty miles upriver from the dam, where I also took students, billing it as a sacred place. There the dam has not only obliterated the sacredness of the confluence by making it unrecognizable, but it has also obliterated history and memory, with the flooding of Cherokee burial grounds and villages. Various historical markers, along with the Sequoyah Museum, dedicated to the Cherokee who invented a syllabary for his tribe, are attempts to salvage that history, to remind us of what has been lost, but they do not replace the vision that we would have of the two rivers coming together, their currents mingling at a sacred and strategic place. Tellico Dam buried a lot of things, and finding where the river revived itself gave me an odd kind of hope based on the secrecy of the place, the intimacy of the deer's call, and even the trash, the toy, something that the renewed current had captured and held like a trophy. Finding Tellico's liminal zone got me hooked on these missions and led me farther afield that summer, fueling my motivation for a larger scale trip the following summer.

ROAD TRIP OF RIVERS

CHAPTER 4

THE CONCEPT

In retrospect the summer of 2008 was not the best time for a road trip across America. Gas was on its way up to five dollars a gallon; the country, on the eve of a contentious election fraught with racial and economic anxieties, seemed on the verge of a not-so-civil war; and driving long distance implicated you in global warming, pollution, warmongering, and the senseless murder of billions of insects. I fretted, sure, particularly about the expense, but I went anyway. Number one reason: it seemed an interesting time to travel the country, a historic time, if you will, when we were on the verge of a dramatic change in leadership. Number two: I wondered if this would be the end of an era, if cross-country meanders might become cost prohibitive for anyone but the wealthy leisure class. Finally, I had a list of rivers that I hoped would lead me upstream to obscure quadrants of mystery and revelation.

That summer I was systematic. I mapped out an itinerary that would cover twenty states and a sampling of twelve rivers in a geographic lasso around the continental United States. I consulted maps and made lists of rivers and campsites. I glanced at other travelers' descriptions and photos to get ideas about what might loom in my windshield, what might come across my bow. Doing a little research—but not too much—before a trip like this is part of the fun. You realize and really hope that no matter how much you prepare, you'll be surprised and challenged, that something or someone will force you to think in new ways about your journey without inflicting permanent psychological, physical, or vehicular damage.

William Least Heat-Moon embarked on an epic loop around the perimeter of the United States in the late 1970s. In his van, which he called Ghost Dancing, he stayed off the interstates and explored the roads that maps once designated with blue, those that meandered through places of enlightenment and unexpected beauty, away from the drudgery of consumerism and commerce to out-of-the-way places where gritty folks would reveal to him the details of their lives. It is a remarkable journey and an amazing book, *Blue Highways*. Unlike Heat-Moon, who was in his thirties when he made his trip and had just been fired from his teaching job and ejected from his marriage, I had just turned fifty, I was happily married, and I'd just attained the status of tenured professor. Healthy and more or less sane, I hadn't anything to escape from, really, just an itch to take a long trip on the roads and rivers of America and to return to the West, a landscape I ached to see again, after a period of living there in the 1980s.

My list of rivers spread across the continent from New England to the upper Midwest to the northern Rockies and beyond to California, returning to East Tennessee on a route that would take me through Nevada, southern Colorado, Texas, and Missouri. I chose rivers, first of all, with names that evoke something of significance, literary or historical: the Gauley in West Virginia, which I'd heard was the fiercest river in the East; the Connecticut River in Massachusetts, across which Puritan Mary Rowlandson, a captive of Wampanoag leader King Philip, waded on her way toward redemption/ransom (I teach Rowlandson's narrative in my introduction to literature class); the Tippecanoe in Indiana, backed up by Lake Shafer, which features an amusement park at a place called Indiana Beach; the Cheyenne in North Dakota, which skirts a reservation and pours into the dammed Missouri River; the Columbia above the Grand Coulee Dam (which Woody Guthrie sang about) near the Canadian border; the King's River near Yosemite National Park in California; the Dolores in southern Colorado, which cantankerous Edward Abbey had rafted before its damming and the formation of McPhee Reservoir; the Brazos in Texas, which John Graves canoed and wrote eloquently about in 1960 just before it was dammed; the James in Missouri, flowing into Table Rock Lake, which borders Mark Twain National Forest and an Ozark Mountain tourist town that goes by the name of Branson.

Going alone—without a friend, canine or human—would give me maximum flexibility. I would be responsible only for myself, and I could complain only to myself if I got too hot or too tired or forgot to bring the water jug. Cautiously I revealed my plans to family and friends. My wife liked the river list and the idea behind the trip, and being accustomed to the two other long river trips I'd taken—down the Tennessee and the Cumberland—she was reassured by the fact that I'd have a car and wouldn't be isolated on river banks in places that evoked James Dickey's *Deliverance*. Nearer to embarkation time, I got the idea that she fly out and meet me in California wine country at a designated date. I was a bit nervous about smoothly pulling off this rendezvous, but Julie worked out the logistics involving her flight, the hotel, and what wineries we might visit, so that such a break in my trip—which I had planned as six weeks— would be a welcome respite from my own cardboard-box cooking and sleeping on the ground, sometimes a ways from civilization. Like Huck Finn, I longed to "light out for the territory," but I didn't mind a mattress, pillow, tables, chairs, plates, and the clink of a glass every now and again.

My mother organized a fiftieth birthday party for me in Murray, Kentucky, where I grew up, and persuaded my brother, his wife, two daughters, one of his two sons, my sister, my aunt, and my wife to each buy five gifts for me that I would use for the trip. I think it was supposed to add up to fifty. They put the stuff in a large blue plastic tub and after dinner watched me open everything from bandannas to polyester blend shirts to six-packs of Vienna Sausages. My nieces, Mady and Libby, gave me a summer's supply of bubble gum, a bag of gummy worms that looked like fishing lures, and a metal container of sushi Band-Aids with colorful drawings of raw fish on them. (If you hurt yourself, as I would, these were sure to cheer you up.) I already had much of what I needed (camping stuff, boat stuff) from previous adventures, but this generous outpouring from my family provided me with essentials I lacked as well as things I didn't know I needed. Julie gave me organic soap made from hemp, organic insect repellent that didn't stink, and a headlamp with a bulb bright enough to blind a grizzly bear a mile away or to divert a herd of stampeding longhorns. In the frenzy of opening gifts, I was able to forget my rather advanced age until my brother, a urologist, raised his glass for a toast. He said some nice things that put me in a good light, quoted Tommy Smothers ("Mom always liked you best"), and gave me the ominous news that while I was still the youngest of three siblings that I was, at fifty, no longer a baby. This I would sometimes forget on my little trip when it would have behooved me to act my age.

I bought my usual array of noodles, rice mixes, and canned goods to mix together for concoctions that would fuel my adventures. Before Huck lit out for the territory, breaking free of the cabin where his psychotic father had imprisoned him, he took a sack of cornmeal, coffee, sugar, a jug of whiskey, a skillet, a dipper, a tin cup, a coffee pot, fishing lines, a gun, and ammunition. I had some version of all of these, except for a gun and ammunition. Unlike Huck, I armed myself with books. Kerouac's *On the Road* went; Henry Miller's critique of America, *The Air-Conditioned Nightmare*; Heat-Moon's *Blue Highways*; Graves's *Goodbye to a River*; Gretel Ehrlich's *The Solace of Open Spaces*, about her time as a sheepherder in Wyoming; Abbey's *Down the River*, a collection of essays; and Lewis and Clark's journals. I also packed a *Reader's Digest* book about weather, which had photographs of the kinds of clouds that forebode catastrophe for people living out of their cars. Once I had everything I needed and was prepared to leave wife and home for six weeks in the bloom of late spring, I began to ponder once again why I was doing this. What was it about? What would I gain? What would I learn? Most important, would it be fun? This was a question Julie would ask me quite a bit as I was en route. Would it lead to some kind of enlightenment or make me a better person? Would I come home disgusted with the country? Worn out? Bored? I was not easily bored, and I was sure I'd stay interested as long as I could keep moving. I could entertain myself. But I did worry about what I would have to show people after such an expensive and time-consuming undertaking.

Aside from my adventures on the rivers themselves, plenty of memorable road stories emerged from the trip. I happened to be wandering through the upper Midwest during one of the worst floods in a hundred years, my stop in Wisconsin punctuated by a one-time hotel stay and my agonizing decision to help people fill sandbags to shore up a failing dam instead of staying in my room to watch my favorite baseball team, the Chicago Cubs, win a rare game (that year). In the far West, the landscape combusted, with more than eight hundred wildfires burning in California. I spent one day traveling through smoke thicker and more widespread than I'd ever seen, the landscape like a tinderbox waiting to explode. Campgrounds were great places to observe homo sapiens outside its natural habitat of couch/kitchen/garage/yard/office, the best and worst of Americans on display; in Folsom, at a state park, a romantic conflict threatened to escalate into a maiming or a murder, one of many sleepless nights I spent on the road. I nearly drowned in a river I added to my list on a whim, surviving out of sheer luck and the kindness of strangers. It wasn't my first road trip across the country, not by a long shot, but in scope and variety it exceeded all the others, and, by the end of it, I was never so glad to see the bridge over the Mississippi near Cairo, Illinois, the pathway back to western Kentucky; nor, at the same time, was I ever so sorry to see the end of such a glorious trip.

In the following chapters are some low- and highlights from that trip. I learned from rivers and liminal zones as different in personality and temperament as the people who boated, fished, and camped alongside them.

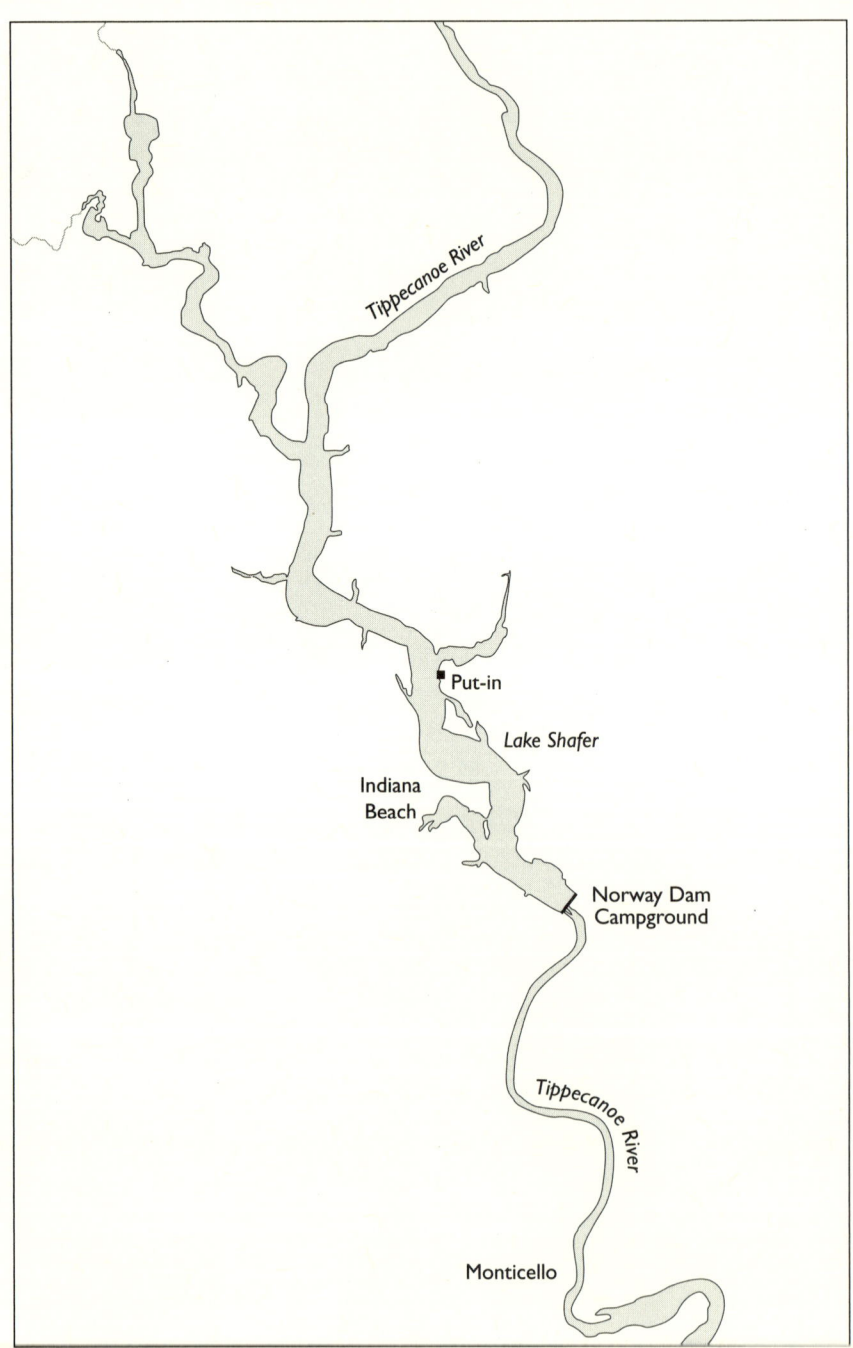

Shafer Lake/Tippecanoe River

CHAPTER 5

EASY WATER: THE TIPPECANOE AND THE JAMES

The toughest boatmen were the French Canadian voyageurs, who not only paddled themselves downstream on rivers that no European had even seen, but paddled and poled and roped their way back *upstream* from where they started. They got out when they had to and portaged across land, their backs bent under ninety-pound bundles of beaver pelts and sometimes freighted with the blubber of a bourgeois or gentleman, whose feet the voyageurs had to keep dry in the marshes and rugged terrain they were crossing. This was part of their business. Mostly they worked in what is now Canada, trading with the Iroquois and the Ottawa, but they made their way south to the Mississippi and the Ohio and the great rivers' tributaries, some of which I would sample in my loop around the continental United States. In this chapter, I bring together what I call the most acquiescent rivers, those whose current and sustained depth allowed me to paddle the farthest upriver toward or beyond their transitional zones, giving me an inkling, however slight, of what the voyageurs felt going against the current.

By the time I got to Indiana, after sampling the Gauley River in West Virginia and the Connecticut in Massachusetts, I'd experienced the following misfortunes: threw my back out zealously unloading my gear at campsite number one; lost my beloved Apalachicola Riverkeeper cap in the high water of Ohio's Mohican River, a downstream diversion that included an unplanned swim; had my electric beard trimmer stolen from the bathroom of Erving State Park near Turners Falls, Massachusetts, after having cut a deep slice in my thumb, a cooking accident that required a couple of the sushi-themed Band-Aids; and camped at the only occupied site in Voorhies State Park, in New Jersey, encouraged to go there by a ranger who said there had been bear sightings there. I'd seen some things I hadn't anticipated, such as a moose crossing I-90 in Massachusetts and sea lampreys sucking the glass of a display case and fluttering like fleshy banners on their way up a fish ladder on the Connecticut River. But nothing prepared me for the first campsite on the Tippecanoe, below Norway Dam, which backed up the Tippy, as it was referred to by locals, into Shafer Lake.

While the summer of 2007 was marked by drought, the skies of 2008 leaked water with a vengeance, dramatic storms and massive fronts transforming lazy clear streams into powerful brown forces of nature; lakes were engorged, full to the brim, with logs and toys and furniture and wrecked boats and other detritus of leisure floating in the coves. Arriving at Shafer Lake, I saw a spectacle such that I'd never witnessed, water released from below a dam with such force that it looked like the flushing of a thousand fire hydrants, a column of white and amber water rising fifty feet into the air, not really that far from where I planned to camp that first night. I picked the campground below Norway Dam, not because of its dramatic scenery or its lush lawn or its efficient staff, but because its downscale, ramshackle identity contrasted so starkly with Indiana Beach on the lake, a massive amusement park with an RV-dominated campground, facilities that include a Welcome Center and gift shop in a giant parking lot overseen by uniformed staff.

At Norway Dam, run by brothers Will and Harold, I was given what Will called the prime scenic tent site, a spit of sand that looked across a small creek at a wall of sheer gravel, crumbling in the constant drizzle that seemed to hang in the air during my time there. Only a couple of months previous, Harold told me, the campground, with a small settlement of campers and trailers, had been flooded by a sudden release from the dam, wrecking the modest dwellings and making a big mess of the entire campground. There hadn't been adequate warning from the power company. "If it storms bad tonight," Harold let me know, "you can come up here with the rest of us in the basement." "Us" consisted of a community of permanent campers who resided in dwellings that ranged from small trailers with decks built onto them to little campers that could be towed behind a truck. Nothing shiny and new here. Very little grass. It was a piece of lowland below a small dam that held back a big lake. No wonder the price was so much lower than Indiana Beach.

Will finished assembling sections of plumbing pipe to form the goalposts for a game called Hillbilly Golf. The goals had three rungs and were hammered into the ground about twenty paces apart. The other game equipment consisted of pairs of golf balls joined by eight-inch-long pieces of nylon rope. The object was to toss the conjoined balls so that they wrapped around one of the rungs—top rung, three points; middle rung, two; bottom rung, one. Will held a can of beer in one hand as he tossed, and I played with my camera slung around my neck. I had a narrow lead when Jerry and Glenda showed up. Glenda, who stood next to Will, was my teammate, and we fell behind quickly. Will was trying to make plans for the yard sale the next day, asking Jerry repeatedly what time he wanted to go. As Will tossed next to Glenda, he kept making not-so-subtle sexual allusions to the balls, to which she would laugh so hard that she began to cough. Jerry thought it was funny, too. Glenda finally got fed up and said, "Shut up or I'm going to crack your skull with this!" She swung the golf balls Ninja style. More people arrived, including Will and Harold's mother, who had visited Gatlinburg and Pigeon Forge and had a daughter living outside Nashville. She noted the irony of someone from Tennessee having to come to Indiana to find out about a game called Hillbilly Golf. She told "William" to stop it when he started up with the innuendos again. I asked Jerry if the fishing was any good on the lake. Jerry didn't fish. He explained that he worked on his trailer and drank beer.

Retiring early to my scenic beachhead, I set up my tent and waited out the rain in my car, hoping the big storm to the north would bypass us. Two boys with small umbrellas fitted to their heads like hats stomped down the creek in the rain, throwing rocks at the geese. A man who resembled the rock star Joe Walsh, if Walsh had been a bodybuilder, hunted down scant weeds around the trailer with an electric weed eater. His girlfriend, who I imagined was the mother of the boys, played angry country music, women singing about love gone wrong. After she gathered the wood and stacked it, Joe Walsh poured lighter fluid on it and lit it. They stood in the glow of the fire, him behind her, his arms around her. Later Harold cranked up the music in the pavilion, the worst of classic rock. He sat there alone at the projector for hours, and the boys joined him for a while, but no movie was shown, just the over-orchestrated music and some kind of advertisements rolling across the screen. I liked the place, but it was so melancholy, soggy, and smelly that I left early the next morning for a campground upriver, at a state park.

The culture next to this moving water was markedly different from Norway Campground, below the dam, and also a world away from Indiana Beach. A deer stood near the gate to the campground, as if posing, and I drove over a mile through the woods to my site in a meadow just a few yards from a boat ramp. My neighbors, a couple, were lingering over breakfast under a nylon canopy—eggs and bacon and toast and jam, it looked like. They had been camping there three days, they said, but, no, they hadn't been on the river yet. I wondered what they were waiting for. I wasted no time unloading the boat and heading down to the small concrete ramp, where two couples were backing in their small fishing boats. Like me, they planned on heading upriver and floating back down. The river was about forty feet wide, the current slow but strong, and I thought I could make it upriver a mile or so at least. I started out ahead of the two fishing boats, as they were loading up beer and food. Mainly beer. I told myself I'd paddle upriver until they caught up with me. As long as I kept paddling and stayed near the bank, I made decent progress. This was the strongest current I'd worked against for a sustained period. At one bend, the river narrowed to twenty-five or thirty feet, and I had to work hard to keep my bow upstream. If I focused too intently on the current as it sluiced against my hull, I found that it had a mesmerizing, disorienting effect. It was better to focus on my goal, the inside of the bend up ahead, where the river widened and the current slackened.

  I passed a rusted-out car on the bank, and it made me think of gangsters, of Al Capone, who used to vacation on Lake Shafer, and of the kind of business dealings that ended at remote places like this, far from where an abandoned Packard with bodies inside might be discovered. Another half mile I paddled, not pausing, quickening the pace at constricting bends and shallow places. There was plenty of water here; one of the fishermen had told me at the ramp that his son had a boat like mine and had been paddling it in the backyard recently. I reached some small houses built just a few feet from the water's edge, vacation cabins with screened porches, all of them empty, paint peeling, screens rusted and torn, no one around. Still the fishermen hadn't shown up, and I was tiring, and though the river remained fairly straight, the width at around forty feet, the current seemed to be strengthening, that or I was getting weaker in my resolve to continue, the lure of the lazy float downstream becoming too much to resist. On this part of the river, I wasn't heading toward a liminal zone. I lacked a tangible goal here; I had no destination, no business deal to conclude, as did the gangsters who deposited the old car on the bank or the voyageurs headed upriver to a trading post to negotiate prices for their furs. So I surrendered and sat with my paddle athwart, revisiting the same dark forests and landmarks I'd been able only to glance at as I had fought against the current. The farther I floated back toward the ramp, the more I worried about the couples fishing and drinking beer. Finally, not three hundred yards upriver from the ramp, I spotted them, their little twenty-horsepower outboards quiet as they sat in the easy water next to the mouth of a creek. They hadn't caught a thing and didn't appear to be doing anything but smoking, drinking, and enjoying their surroundings.

The next morning, around 7:00, I was on Capone's lake, above Norway Dam. For four hours I paddled the flat water of Lake Shafer, where the houses were fairly modest, the lots so narrow you could survey all your neighbor's business, were you so inclined. It seemed as if a third of the properties were for sale. I passed a couple of men leaning out over their deck railings, smoking. One, wearing sunglasses, grinned and waved. There were no boats out. At the mouth of a cove stood a green British Petroleum sign, advertising a marina. Under it stood a statuesque woman in a gray crocheted swimsuit, skimpy and quite revealing. I stopped and took photographs of her in the glow of the new day. She didn't mind. She was made of plastic. From across the lake she might arouse a lonely gangster, his vision a bit fuzzy from late night conspiracies. I wondered whose idea this could have been. Certainly a man's. Did he have a wife? Was she jealous of the plastic woman, of her prominent place on the point, of her height, over six feet? I wondered if the wife had a good sense of humor and knew what drove business into the cove.

After four hours, I came upon the liminal zone under a small highway bridge where a group of men had gathered to fish. Cottonwood seed pods swirled in circles on the surface. I paddled under the bridge and the current began moving the seed pods toward me. The channel narrowed to thirty feet, and, after a half mile, thick grass covered each bank. Here, the Tippy didn't let me go far, the river more narrow between the grassy banks, hard charging and shallow. I accepted the rejection and nodded at the fishermen a second time as I was pushed to the flat water.

"Is this where the Tippy comes in?" I asked.

The oldest of the men said, "It's a branch of it."

The James River

Table Rock Lake, in the Ozarks, is one of a cluster of lakes in a system of dammed rivers in southwest Missouri, northeast Oklahoma, and northwest

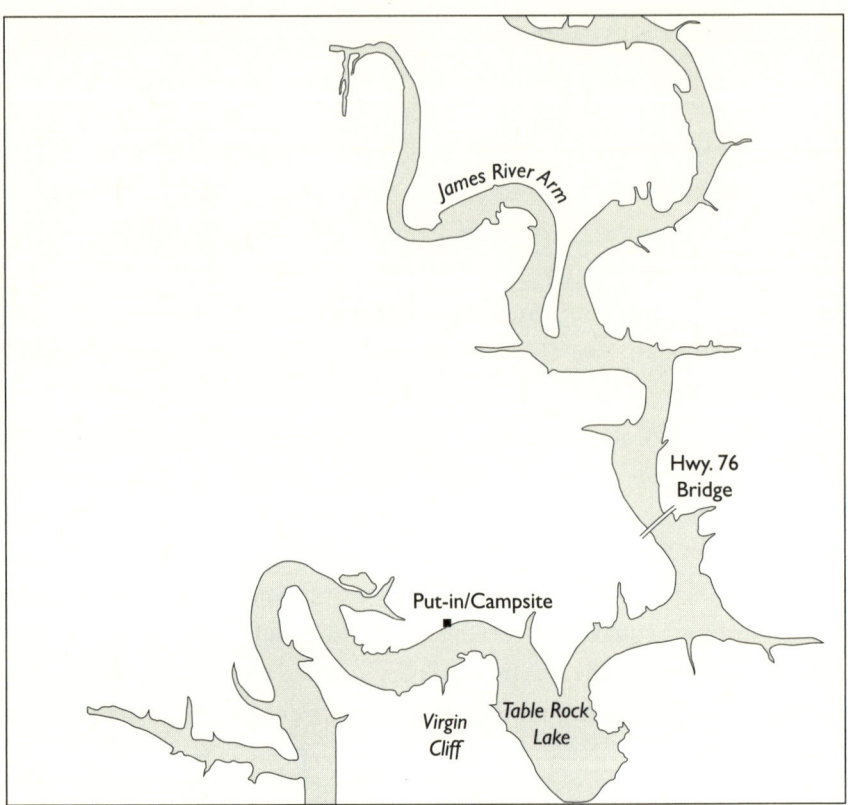

Table Rock Lake/James River

Arkansas. The White River feeds Table Rock from Arkansas, as does the Kings River, but it was the James, coming down from the north, from Galena, Missouri, that I chose, because it seemed the largest of the lake's "arms." I targeted the jumping-off place for the James River transitional zone near Cape Fair, where a bridge acknowledged the connection between lake and river with this sign: "TABLE ROCK LAKE/JAMES RIVER ARM." From Cape Fair I descended into a deep hollow on a narrow shady road to reach the Corps of Engineers campground spread across a couple of forested slopes. I picked a site not far from the entrance, just across the road from the RV of a camp host. I failed to notice, somehow, that a dumpster the size of a school bus lurked just a few yards upwind of my site. Below me trees and picnic tables stood half-submerged in the risen waters. The man at the office said there had been six inches of rain in the last two days, the latest installment of intermittent flooding that spring and early summer. The lake, he said, was eight feet above flood stage. He mentioned that a boy had been bitten by a water moccasin that had fallen into a fishing boat from a tree. "His father took him all the way to Springfield," he said, shaking his head, "when Branson has a perfectly good hospital." As one who likes to boat under trees, in the shade, I appreciated the cautionary tale, and I filed away the not-so-understated Bransonian boosterism.

Into Branson I went, after having popped up my tent and set my boat down to rest on the soggy ground. I was camped above Table Rock Dam; Branson lay below it on upper Lake Taneycomo, a section of the White River stocked by a trout fishery. Farther down, the White is pooled again by Power Site Dam, finished in 1913, the first of the eight that would tame the White. Eventually, the 760-mile long White makes its way across Arkansas to the Mississippi River way down south, across the river from Gunnison, Mississippi. Where I was, the White was obliterated under a flood on top of a flood, its once-clear water and gravelly beds gone, its current artificially revived in the tail waters below dams. The dams, created for flood control and power generation, profitably complemented the culture that rose up around the lakes: motorized recreation, houseboats and cabin cruisers, big boat docks and marinas, bass fishing, and water skiing.

I parked at the lakefront at Branson Landing in a lot that extended a mile down the lake, where it ended at a campground full to capacity. The Bass Pro Shop (White River "Outpost") greeted me like a monument at the entrance to the riverside mall, which was lined with stores that sold everything from fishing gear to oversized clothing, from bubble bath to diamonds. You could dine on Italian, Irish, Mexican, or coastal. There was ice cream, pizza, fudge, coffee, subs, and gyros. The Baldknobbers Jamboree poster caught my eye. On it were head shots of the cast members, set out in neat symmetrical rows. On the bottom rung were two beauties, a redhead and a blonde, and, at the top, three beautiful men with spiky hair cut short, and white, white teeth. They would try their gol-darndest to entertain you; you'd trust them to do their best. In between these two rows of perfection were three men, two of them toothless, their features distorted by studied facial contortion, and a third who grinned with teeth that protruded from his mouth over his lower lip, an orthodontist's nightmare. This act, founded by the four Mabe brothers in 1959, was the oldest in Branson, and, in a way, it seemed to personify Branson-ness, or at least a version of hillbilly Ozark that would confirm and reinforce visitors' preconceived notions about mountain folks. From other posters, I found out that Andy Williams was alive and still crooning, alongside Ann-Margret, and from the poster it looked as if neither had been marked by the passage of fifty-odd years beyond their prime; in fact, they seemed better looking, more eager and vivacious, than I remembered them on my parents' flickering black-and-white TV in the 1960s. I learned that Branson wasn't really a country-music town at all, that it was more of a cover-band town. Bob Anderson, who on his poster posed in a tux while getting out of a black car, "delivers the big sounds of Tom Jones, Neil Diamond, Tony Bennett and more in a totally cool show featuring the best of the 'rat pack'!" The Beatles were in Branson! Well, no, actually it was the "Beatles Experience," presented by Louise Harrison, George's sister. "Noah: the Musical," complete with "300 live and animatronic animals," re-created the ancient flood that occurred long before Table Rock's dam. Kirby and Bambi, a beautiful human couple, frolicked with a white tiger and a leopard as part of their magic show. This last one might have been worth the trouble of sticking around for that evening, but the ticket price, thirty-five dollars, was a little off-putting for a cheapskate like me. (That's two night's lodging!) I wondered if Johnny Cash had ever come here, or better yet, now that Cash had died, if Bob Anderson would add to his repertoire "Big River" and "Ring of Fire."

Heat glared from the broad white sidewalks crowded with strolling consumers, who seemed mostly happy and hungry. Business was good in Branson. Down by the river was a sleek cruiser worthy of an oceanic crossing, and chugging down the lake was a canopied pontoon, its broadside lettering—"RIDE THE DUCK"—halfway underwater, freighted, as it seemed, with an overload of tourists. The onshore centerpiece was a waterworks such that I, in my sheltered life, had never seen. I stood before it and gawked, Neanderthal-like, for thirty minutes in full sun. It shot water into the air, of course, but it also blew orange flames from a row of ten or so rust-brown smokestacks, all of this in syncopation with music coming from somewhere. The fountain seemed to be mocking my mockery of it, Joe Walsh's "Rocky Mountain Way" blaring from the hidden speakers. If you don't like it here, the waterworks said, go back across Kansas. Okay, okay, I replied to the waterworks god, I'll try to keep an open mind. I'll try not to judge, lest I be judged for my judgment. So, let me try this again, without irony. At the waterfront is "a vibrant town square terracing down to the $7.5 million spectacular water attraction that features the first-ever merging of water, fire, light, and music." The fountain was built by the same folks who created such "features" as the Bellagio in Las Vegas. The Branson geysers shot 120 feet into the air, and the proper term for what I stupidly thought were smokestacks: "fire cannons."

I tore myself away from the cooling mist of the fountain and reconnoitered the rest of the mall. Inside one store window, two boys stood frozen like mannequins, waiting for passersby to notice them. I did an open-mouthed double take that made them blow their cover, but when I aimed my camera at them, they froze back into their counterfeit poses. On the way back, I saw something else that made my jaw drop: a cedar canoe standing on its bow in the corner of a display window. It was maybe eight feet long, with a high rounded bow and stern, an ornate design patterned after Native American vessels. I'm fairly certain it was manufactured for this purpose—as a prop for merchandise—and I have no doubt that it had never touched the water, but it had been scraped and "distressed" by someone to make it look used, as if an Osage brave had paddled it up Lake Taneycomo for a lunch of crab cakes at the Fish House. Worse crimes had been committed against canoes and canoeing, and I had been a perpetrator of some of the most heinous scrapings in the history of river travel, but to display an imitation of such a noble craft in the window of a trendy clothing store was heresy to a river worshiper. There was no price tag on it, and I hadn't the heart to go in and ask about it. In the Bass Pro Shop existed a parallel universe of "Nature": waterfall, forest, pond, live fish, and dead animals dominated the merchandise in a store the size of a basketball court. A woman cleaned the glass of a fish tank that fronted an ecosystem extending up and back to include a waterfall, waterfowl suspended from wires, bobcats, raccoons, coyotes, and elks frozen in their wildness. Fish leaped from a pile of camo pants. A wild boar poked its nose through a rack of sweatshirts. A bear reared up and raged at bass boats with motors big enough to shoot you from one cove to the next without breaking a sweat. I walked through the whole thing, as if doing inventory, and walked back out the other side into the parking lot without having spent one cent, though I saw things I might have used.

As if to avenge my high-handedness, the travel gods confused me on my way home to camp. I wandered through a maze of four lanes and winding narrow roads that ran along the tops of wooded ridges before I made it back that evening, when the door had been shut to the day's furnace. The stench from the "dumb-ster" had intensified during the day. I held my breath and walked up to it. Someone had left the lid open, and the detritus of Fourth of July weekend rose without impediment into the damp air. I swung the door down, and the slamming of metal on metal clanged across the flooded grounds.

In the morning I tiptoed over the mud, kayak on my shoulder, to the swampy bank, scanning the ground for snakes. I set the boat down in the water beside something plastic that floated half out of the water, ballooned like some parody of a jellyfish. The air smelled of dead and dying fish. I paddled past the Virgin Cliff, a smile of tan outcropping within the greenery on the far shore, from which a heartbroken Missouri girl had allegedly cast herself. I slogged onward up the lake, under the bridge, past the fork for Flat Creek, an enormous cove. Water stood in fields and hid barbed wire fences. The surface was covered with sticks and scum and plastic containers for the hydration of humans and the motors that propelled them. Four hours later, weary of the big ugly lake, I idled at a boat ramp and chatted with some bank fishers, a man, woman, and two boys, whose parents were calling them from a house a hundred yards up the hill. They had caught nothing. Yes, they were tired of the rain, of the flooding, and there was more to come. A man drove up and screeched to a stop on the mostly submerged boat ramp. One of the fishermen smiled without looking around. When the driver got out, the fisherman said, "That's what kind of boat you need." He pointed at me.

The guy nodded and stared at me, about fifteen feet offshore. I went through an inventory of my boat's advantages. It sounded like I was trying to sell it.

"How much was it?"

When I told them, they looked away. Paying that much for a boat without a motor? At this point on my excursion up the James Arm, I was inclined to agree. I had reached a point on the lake map near a bend where the phrase "James River" was printed alongside the winding flat water I was paddling. It would be perhaps another three miles to reach the liminal zone. No current stirred here. I wouldn't fool myself. It was about noon, and the air was humid, ripe for storms. I had at least four hours of paddling back down the lake. Back at the bridge, I'd nearly finished off a gallon jug of water, and a fisherman casting against a pier nodded at me as I went past.

"How are you?" I said.

"Good."

"How's the fishing?"

"Not good."

He took a long look at me. Old enough to remember the James Arm before Table Rock Dam was built, he had the weathered look of a man who had cast many lines. "You been a long way today," he said. "I saw you this morning when you started out." He seemed a bit concerned, perhaps impressed in the way that one is sometimes impressed at what the demented or deluded manage to accomplish.

I was near delirium from heat and dehydration when I noted Virgin Bluff and veered across the wide lake to a point that I thought hid the cove for my campground. The more I paddled toward it, the less I was convinced that it was my cove. No matter how lost or disoriented, I wanted to avoid, at all costs, paddling too far and having to backtrack. A pontoon with three women launched from a ramp and puttered out into the middle of the lake not far from me. The pilot cut the motor as I gestured to them and approached.

"Do you know where the Corps of Engineers campground is?" I asked, unable to remember its name, Cape Fair.

Long pause. "I think there's a place to camp up there," she said, pointing to an opening at least a mile farther on. She'd mistaken me for a kayaker on an expedition, looking for a place to camp, not some poor sap who'd lost his way back to the place where he had his tent and his car. I looked ahead to where she pointed, not fully convinced. I would die of disappointment if she was wrong and I had to paddle back three miles.

"Thanks," I said. "I'm a little disoriented."

"Wonder why," she said before pulling back on the throttle.

As they motored away, in no hurry, going just fast enough to create their own breeze and leaving a fog of sweet exhaust, I debated what she meant by "wonder why." It seemed a sarcastic reference to my delirium. Here you are, she seemed to imply, in that silly yellow boat without a motor, your brain baking in the sun, drinking warm water from a plastic jug. No wonder you're lost. Don't you know how to have fun?

Apparently I didn't. Apparently, in the process of taking this trip, I had become alienated from the region that spawned me. Something had happened the day before that still puzzled me, even after discussing it with others. It was an incident that left me with an image so unsettling that I considered not even mentioning it, though, to make complete sense of this transitional zone, I have to describe it. On a walking tour of the campground, I noticed a rotund man in shorts and running shoes rolling a tire toward the water's edge near the public boat ramp. At first I thought he might veer to the right or left either to dispose of the tire or to take it with him and salvage for a swing or a planter, and I was thinking, what a good citizen to clean up the campground so that others didn't have to look at and smell the tire. What he did was roll the tire into the lake. There was a man standing beside him in a red bill cap, a spotless neon yellow jet ski parked just a few feet away on the water. I took a hurried photograph of the tableau just to be sure of what I'd seen. It seemed the ultimate in absurdity, to roll a tire into the water in broad daylight, but the brazen, leisurely way he went about it made it clear that he thought that this was acceptable behavior. It raised many questions. Was this a tire he'd found on the shore, and this was his way of disposing of it? Had he brought it here to the campground for the purpose of getting rid of it? I'd seen sections of the Cumberland River in Kentucky that resembled depositories for discarded appliances and automobiles. I'd seen a toilet seat hanging from a tree branch on the lower Cumberland. But never had I witnessed someone engaged in a deliberate act of large-scale littering.

When I showed my colleagues slides from my trip and ended with this one on a tone of wonder and consternation, Drew Crain, a biologist, said this: "Where I grew up in South Carolina," he said, "we used tires to catch catfish." Drew, an affable man with a talent for capturing salamanders with his bare hands (and releasing them), was not so much defending the Missourian as he was giving me another perspective from which to view the act. The catfish, he said, nestle in the tire, which is slit so that it could be pulled from the water with minimum effort. It is an easy way to harvest fish, big ones, he said, and, one might argue, a bit more humane than hook, line, and worm, one that wouldn't require the burning/leaking of gas and oil that many indulge in during their pursuit of fish. Still, the guy looked like a tourist, a transient, not a local who would come back and check the tire for fish, and the location, next to the boat ramp, seemed an unlikely place to harvest catfish in this way.

As disoriented as a catfish caught in a tire, near the end of my kayak excursion, I paddled past houses on the bank that looked unfamiliar, their boat docks askew from the high water. I asked another boatman about the location of the campground. Like the fisherman, he recognized me from earlier. He affirmed that I was at the mouth of the correct cove. "Hard to tell where you are in this high water," he said. That's what the boat woman had meant, I realized: it was no wonder I was disoriented because the flooding had changed the lake, submerged landmarks that guided the way. It also occurred to me that high water would push back a transitional zone, that it might have taken me another four hours of paddling upriver to see physical evidence of the James's current.

I let the bow embed itself into the muddy bank and arose from the boat to drink some more water. I trembled a bit. My hips were stiff, and I moved about massaging the small of my back before shouldering the boat. In effect I was fleeing Branson and Table Rock Lake. The heat, the high water and what it had washed ashore, the crowds and traffic of Branson, all this made me long for something pure, isolated, and undiluted, maybe the Tippecanoe before Tyler came and sacked the Indians, the James back when Mark Twain was a cub pilot on the Mississippi and hadn't yet written *The Adventures of Huckleberry Finn*.

I had a lot to learn about paddling upstream, and I was hungry for it, appreciative of those who had come before me like the voyageurs and modern-day marathoners such as Verlin Kruger, a canoe inventor and builder who paddled one hundred thousand miles in his lifetime. Kruger had made epic trips across oceans and continents, but what impressed me the most was that he'd paddled up the length of the Mississippi. *Up* the Mississippi, a voyage that seemed unthinkable. Kruger was not all fire and blood; he was a guy who thought a lot about maximizing his effort, who built boats that were not only efficient and customized for the particular trip he planned, but also comfortable. Any time you're uncomfortable, he said, you're losing energy. How true.

All this thinking that I'd been doing—deriding Branson, where folks were merely trying to have a good time, paddling until I was delirious from heat and exertion, forging upstream without a tangible objective—this wasn't normal, and I knew it. Nearing the end of my loop around the country, I was beginning to be at peace with not being normal. In a sense my position at places such as Lake Schafer and Taneycomo Lake was, as Belden Lane might say, "betwixt and between." I was apart from the crowds, but also among them. I might make fun of what I observed—the water cannon, the outdoor store with all the dead animals—but I was aware of my own absurdity, how I might look to people who saw me paddling up the arm of a flooded lake. A month earlier, back in Oregon, I tried to fit in, just for a day, and it almost killed me. I tell that story in the following chapter.

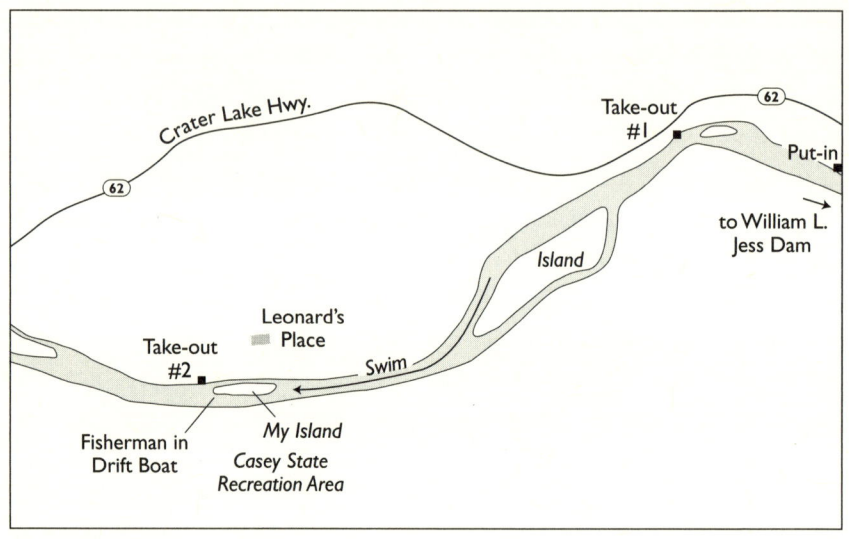

The Rogue River

CHAPTER 6

THE ROGUE'S EMBRACE

"What's there to do in Sisters?" I asked the waitress after I'd settled onto a stool at the counter, where I could watch the cook prepare my breakfast. The waitress, pretty far along in her pregnancy, seemed to be enjoying her job at the Ski Inn, at least for the moment. She mentioned a nearby lake and a museum in Bend. I asked her about the famous Crater Lake, maybe fifty miles south, and she said she'd never been, though she'd like to sometime. Hearing this made me a little sad for her.

The roads to Crater Lake National Park, I discovered, had just been cleared of snow. In July. No rivers fed the lake, which was formed after a volcanic eruption and the subsequent collapse of a twelve thousand-foot-tall volcano. The lake was filled with rainwater. It was, according to the National Park Service, "one of the snowiest places in America." I had no doubt that it was worth seeing, that there was no place like it, but I decided not to camp at Crater Lake. I might visit, but for camping I decided on the ominously named Lost Creek Reservoir, fed by the ominously named Rogue River, which I deemed a safer, more convenient, and softer adventure than driving up to Crater Lake, a wintry place for which I was ill prepared. In short, I chickened out.

My campsite at Joseph H. Stewart State Park afforded a view of the bathhouse, a volleyball court, and many other campers, bivouacked across the mown fields. There was no view of Lost Creek Reservoir, nor of anything like Mount Jefferson, the snowcapped peak I had gazed at the previous night while at Lake Billy Chinook State Park. Here I was camped on a small rise under a big dark Douglas fir, and several yards below me was a campsite impoundment of several tents with at least ten people and one guy, short and thin, with an unusually loud deep voice. Moments after I'd arrived, he was saying something about "dumbasses," and though I don't think he was talking about me, I went back to the office and arranged to move one site over from Little Deep Voice, who resembled the late Frank Zappa—black hair, Fu Manchu mustache. From the new site, I could still understand every word he shouted, including the fact that he could no longer eat almonds, but it didn't sound as if he were inside my head.

A brief hike made me think I was back in the Tennessee Valley. I had descended to the southern tip of the Cascades, where a mix of hardwoods and evergreens stood on hillsides that gently sloped down to an expanse of impounded water that was not as cold, not as clear, and not nearly as blue as Billy Chinook. People caught catfish from Lost Creek Lake. By that evening, a Friday, the campground would fill to capacity, and the air had that Fourth of July sweatiness about it, even though the Fourth was a week away.

I talked to a camper who had a wooden boat with an upturned bow and stern. I'd fished from one of these on the South Platte in Wyoming, when I'd attended the University of Wyoming, but I'd forgotten what they were called. "It's a drift boat," he said. "They used to be called MacKenzies."

I told him I was going to paddle from the lake up the Rogue as far as I could go. He said his brother liked to kayak down the Rogue. "It can get pretty nasty up there," he said.

"What do you mean?"

"It can get really intense," he said.

For some reason, the vague way he was putting it made me think he was talking about people who lived on the upper Rogue. I was thinking *Deliverance* villains. "Do you mean people up there will bother you?"

He squinted at me, trying to understand. "No," he said. "The river is very dangerous."

When I returned to my site, I had neighbors on each side of me. On one side, next to Frank Zappa, were a man, a woman, and a johnboat. He was setting up a tent. Grinned at me and said he'd just gotten it out of the box. On the other side was a guy standing next to his bike. He waved. I walked toward him, and he asked me where I planned to kayak. "Up the Rogue," I said. He said that many liked to paddle the segment below the dam. He had just come from Lake of the Woods.

"That's a long way," I said, thinking he'd ridden from the famous Lake of the Woods in Minnesota. "I've been across country myself, all the way from Tennessee, via New England." He walked toward me a few paces, his head cocked to one side. He had on a cyclist cap, close cropped hair beneath it, his body muscled up from the waist down. He was about my age. He and his wife, who was showering, had *ridden their bikes* from Lake of the Woods in *Oregon,* not that big ugly lake in Minnesota. "Where are you from?" he wanted to know.

"I live in Tennessee," I said, "but I grew up in Kentucky." I always take a certain pride in announcing this, particularly in areas far from home, where some consider my native region exotic, eccentric, maybe a little dangerous. The combination of Kentucky—Daniel Boone, thoroughbreds, whisky—with Tennessee—Davy Crockett, walking horses, whiskey—created in people's minds a heady mix of momentous history, rebellion, and outlaw culture, or at least I liked to think so. What people actually thought about me when I announced my roots in Kentucky and Tennessee, they withheld. The camping cyclist did not.

He twisted up one side of his mouth to say this: "God, you come from where people run their air conditioners all the time and pollute the air. How do people live there? It's horrible. There was nothing going on there until the invention of the air conditioner, and then they started burning all that coal. Miserable."

He went on and on like that, as if he'd written this down and rehearsed it. He was eloquent and unqualifying in his distaste, and I knew not what prompted it, but I was so aghast that all I could do was stand back and stammer every now and then. The diatribe was mainly about the climate but implicated southerners (and easterners, too) for being wasteful and stupid for inhabiting such a place.

I think he ended the first round with something like "How can you stand to live there?"

"It's my home," I said. "You get used to the heat and humidity. Not everybody runs their air conditioner full blast all the time."

"I do like the West," I added. "But back there, that's home."

"Good. I'm glad people back East feel that way," he said. "It keeps it cleaner here."

He told me he had ridden his bike across the Southeast, that it was horrible, horrible. "I was just drenched in sweat all the time," he said. Yes, he'd been to the Smokies. He'd biked the Blue Ridge Parkway. "Not really mountains," he said. "And no relief from the heat. Horrible. Yuck!"

"Seemed to work fine for the native peoples who lived there for centuries," I said somewhat lamely. "I don't think they had air conditioners."

He conceded this. I had a point. He would not rise to the bait of attacking the Cherokee, Seminole, or Creek peoples. "Well," he said, "New England was not bad. Maine was okay."

I think he figured out that he was irritating me beyond tolerance because I started to edge away toward my picnic table, too flustered to mount a reasoned argument. How do you defend the weather and culture of half a continent? He switched gears: where he thought I should go next. Crater Lake. I had to go to Crater Lake. He was only here, at Lost Creek Lake, because it was the end of his bike trip. It was also "a big ugly lake."

Of Crater Lake he said, "It's the bluest blue you'll ever see. And you know, of course, that the blue is merely refracted from the sky. The water is so clear. A French helicopter pilot crashed onto the surface a while back because the blue sky and blue water disoriented him." He added, "You can't put your boat on the lake."

That settled it. I would not go to a lake that disallowed boats. Besides, people had promoted the place to the point that actually visiting it would surely precipitate a collapse similar to the geological event that created the crater. It couldn't possibly live up to its billing, and it would take a day to get there, two days to see it, and a day to get out. I liked Oregon, but not that much.

"Thanks for the tip," I said. "I'll think about it." Then I turned away from the guy to cook some rice.

His wife returned from the shower and began to work on the tire of her bike. John, the guy, came over with a map and showed me his place on Lake of the Woods. I should visit there, he said. He had a paddle boat that would go faster than any kayak. Did I like sailing? He would take me sailing.

Now he was like a little kid wanting me to join his club. "I'll think about it," I said. "I'm moving on toward California. Going to meet my wife in Napa next week."

"Then you should see the redwoods on the way. You'll certainly wax poetic there." "None of this is old growth," he said, extending his arm to include the entire park. "It's all been cut over. That's Doug fir," he said, on a first-name basis with the tree. "Fast growing, probably only twenty years old."

He wrote his name and number on a piece of newspaper. I feared he would return with more information, more opinions, but I think that his wife, who never even looked my way, may have suggested to him that he'd said enough.

Why do people assume credit for the place where they happened to have popped into the world, the land from which they sprouted? As if they had anything to do with it. You don't hear people brag so strenuously about places they've moved to, a choice for which one might deserve credit or blame. It's the birth region that evokes emotions, arguments, or disputed criteria for the quality of life. What's the happiest place in the world? Denmark, according to a recent poll. The unhappiest? Zimbabwe. Obvious criteria about standard of living and violence. But this stuff that Oregon John was talking about, it implicated more than politics and climate. What he was saying came down to this: I'm better than you because of where I live. I'm beautiful because I live among beautiful things. Plenty of southerners I know would have got up into the face of Oregon John for sharing such opinions. And, to be fair, a good amount of southerners are just as chauvinistic about the tragic superiority of their birth region.

What was ironic about my irritation was that I'd been thinking similarly since I entered Montana, maybe even sooner, like in Minnesota. Why wasn't I living out here? How come westerners don't trash the landscape? For over a thousand miles, I think I'd seen one cigarette butt flicked from a car window, and that from a car with a nonwestern license plate. Back home it happened every five miles. No one had tailgated me. Why weren't there billboards proclaiming the imminent conclusion of time and the universe? So while I might have been inclined to agree with some of what Oregon John had said, his delivery and his extremism put me into a position of defending a place about which I was critical myself. The difference? I was from there, he was not. I didn't want to hear someone say that I lived in an "uninhabitable" place. It made me feel stupid.

Before dawn I put in on Lost Creek Lake and paddled under the Highway 62 Bridge toward a narrowing stretch of calm water where the Rogue would awaken itself. The sky was overcast, and breezes fluttered the trees on the bluffs around me, mostly evergreen, some Manzanita, a dogwood-sized tree with a smooth reddish bark. Thunder sounded down the lake, where a patch of sky darkened. An osprey launched itself from a treetop, clutching a three-foot limb. He dropped it over the water, and it fell in slow motion, the splash louder than I expected. When I got to the point where the rocks crept out from the shoreline and islands emerged, the thunder closed in. No big claps, just a shaking of the cosmic dice. I pulled onto the bank, sat on a flat rock, and pulled the poncho over my head. Rain ensued, light, pleasant in the warmth, a straight down shower that freckled the Rogue. Fifteen minutes later it stopped. I paddled three hundred yards to where the current poured over the rocks. Up the Rogue, where I couldn't go, the forest shadowed everything. It was like the dark opening of a cave, this transitional zone, the threshold to the place that the drift boat guy called "intense."

Oregon John and his wife had pedaled back toward paradise by the time I returned to camp, about mid-morning. While I refused to follow his advice to visit Crater Lake, I yearned to go along with the crowds who paddled downstream on the Rogue, who screamed and laughed when sprayed by cold, clean water. I drove to the tailwaters of the William L. Jess Dam, where fifty or so people stood on the riprap, casting lines into the moving water. There was a hatchery upstream. Barefoot teenagers inflated rafts with foot pumps. A girl told me they were going to float about ten miles.

"How rough is it?" I asked.

"It's not bad, but you'll get wet."

I had a boat on top of my car. The Rogue was running right there. It was speaking to me. I drove up and down the highway that ran alongside the river, scouting as well as you can driving a car. At one spot, I parked fifty feet above the river where people in rafts ran through a turbulent bend. On the far side of the river, a man worked the oars of a drift boat in an eddy behind two tent-sized rocks. A woman in the bow and another in the stern *stood* and cranked fly rods back and forth, their lines undulating toward the shore. It seemed effortless, his keeping the boat steady when so much big water was rushing all around him. On the near side, in the main current, a man and woman bounced through the waves on an inflatable kayak, the woman in blue jeans and boots, reclined, the guy stroking leisurely with a plastic paddle that looked like a toy. Another guy shot past in a glorified inner tube. Several fishermen stood on the bank, drinking beer and casting into the water. Everyone was having fun, and here I was just watching. I found a campground three miles downriver where I could take out, and I parked upriver from the bend, where the rafting companies were dumping off customers in six-person rafts. In front of me, a muscular young man was fishing with his girlfriend, a tall, striking woman who had her jeans rolled up, water rushing above her ankles.

The guy walked from the beach to his pickup while I was changing into my rubber boat shoes and putting my wallet and keys in a dry bag. I left my cell phone in the car. I put on my life jacket. Then I made the first of several bad decisions. Decided not to use the spray skirt, the neoprene thing that fits over the cockpit of my kayak and keeps water from splashing in. Weren't all these other people getting wet and having fun? Why wear the hot skirt?

The fisherwoman waded up onto the beach. The water was so cold, she said, that it had numbed her legs. The fisherman looked my way. I said I hadn't seen anyone catch a fish all day.

"I caught a pretty good one," he said. He lifted a seventeen-pound Kokanee salmon from his cooler and cradled its belly with both hands. Yes, it was good to eat, he said, returning it to the cooler. He put in a fresh wad of snuff into his mouth and rejoined his girlfriend on the beach.

"Have fun," he shouted as I launched the kayak a few minutes behind the last of the rafts. Within moments of embarking, I realized that I was in trouble. The waves were much bigger than they looked from the road. (Imagine that.) The hull of my kayak bounced heavily—BOOM! BOOM!—and buckets of water splashed over the bow. When I reached the approach to the bend, the waves got higher, two to three feet, and I struggled to maintain a line with the current as it charged toward the wall of rocks where the beer-drinking fishermen stood. Up ahead, the drift boat was still there. Waves came from all sides as I entered the chute at the apex of the bend, the boat half full of water, unsteerable as I approached the passage, a ten-foot gap between rocks. I went through backwards and managed to eddy out at the shoreline. I turned the boat upside-down and set it back down on the ground. A few of the fishermen glanced my way and then went back to fishing.

I considered taking out here. There was a path from the river to the road above, and I could walk down the road to my car in fifteen minutes. I could say I'd paddled the Rogue, without mentioning that it was only a quarter mile, and I'd have a good story about how foolish I'd been to try the Rogue in a flat-water boat, without the spray skirt. I had gotten a little wet, and I was winded from paddling a lot, to little effect, and from nearly capsizing in front of dozens of people who did not want my flailing arms and legs in water they were fishing.

The stupid part of my brain began to take over. Surely, it told the tiny smarter part, that was by far the worst of the river. I should go on at least a couple of miles. I can't just admit defeat here. To think of paddling a river as a matter of victory and defeat was stupid to begin with, but as a competitive soul, I let it creep into my reasoning. In short, I relaunched.

"This is the worst of it, right?" I said to a fisherman on the bank as I floated past him.

His reply? No reply. What could he say? The blank look he returned worried me because I was past the point of turning back. What he left unsaid was this: no, that was not the worst of it.

The river widened—I remember that. I took on water in the next minute or so, enough that the boat wallowed in the increasing waves. Up ahead the water was just as rough if not rougher than what I'd passed through, only now it continued, splitting an island up ahead. I capsized without much ceremony, knowing that it was inevitable. The shock of the cold water was electrical, and the animal part of my brain kicked in. Swim, it said; get out of this.

Because I knew that I would capsize, I held onto the swamped boat and the paddle. To say that the river carried me forward would not convey the nature of my swim. I had a hold on the boat, true, but much of the time, my head was below the surface, and I fought to stay in an upright position. The Rogue invaded my nose, mouth, lungs. I coughed. I sputtered. I looked to each bank, about an equal distance away of thirty yards at least. I noted the absence of my prescription sunglasses, torn from my face in spite of the band that was supposed to hold them. It seemed that I was fighting with the water for the length of a football field, and I began to perceive that if I didn't somehow stop myself at the point of the island, which was coming quickly, that the river would put me in a truly serious predicament. I had enough time to think out there in the waves, enough time to realize that the Rogue was not to be trifled with, that I had trifled with it, and that I was now paying a serious penalty. I didn't think I would drown, but being in that cold, cold water and having some trouble keeping my head above it certainly introduced me to the distinct possibility of perishing.

As the island approached, I got my feet up in front of me and slid into the gravel bar. The current was so strong that it tried to pull me on around the left flank of the island, as if to say, "I'm not quite finished with you." With my left hand, I held onto the boat, which contained about thirty gallons of water, and with my right hand grabbed at the gravel, but as I floundered sideways with my body, I could not hold the paddle. It disappeared around the side of the island toward the drift boat with the two women standing in it, near the downstream tip of the island. I turned the boat over and stood shivering next to it. Gathering myself. Assessing. I had a boat but no paddle. And even with the paddle, the boat was nearly useless. The drift-boat fisherman and his clients could not see me, and they certainly couldn't hear me over the roar of the river if I shouted at them, which was the last thing I wanted to do. A couple of kids floated past on inflatable kayaks. They whooped and they hollered as they rocked in the waves. Barely glanced my way. Their father was in a separate kayak. Kids. Not over ten years old, maybe younger. Perhaps they wondered why I had stopped here. It would be hard to do even if you tried, the current was so strong, the water so deep on both sides. To the right, where I needed to go, it would be a hard, lethal swim of twenty yards, the bank offering no easy access, especially with such a current. I could jump in and swim on downstream, sure, but I could not stand even to think of immersing myself in this water again. I was still shaking, my teeth chattering. I had no idea what would await me downstream. It could be more of the same. Or worse. I was a living example of what happens to fools who embark on a river without scouting it. I knew not what awaited me.

I walked the length of the island, about fifty feet. Thorns tore at my flesh. I stumbled over brush and driftwood and tilting rocks. A tiny snake slithered out from under a log. There was something clinging to my wet neck that I slapped at. I pulled it off, the strap that was supposed to hold my sunglasses. The point of the island was thick with willows, and without my glasses I could not get a clear idea of the downstream situation. I scanned along the left flank of the island as best I could, hoping that the paddle had washed into an eddy. No such luck. I kept walking around, trying to get dry. Peeled off my shirt and stood at the upstream tip of the island, wishing that the sun would stay out longer than ten seconds. I stared up at the sky like a Neanderthal just emerged from his cave.

I did not have my cell phone, but I knew that if I waved my arms in distress at one of the passing drift boats or rafts that they would summon the rescue squad. The rafts and drift boaters could not stop here and pick me up. No one with any sense would be on this island. What people thought as they passed, I do not know. I averted my eyes. Just a couple of days previous I had read with derision an article about two girls, Abby Flantz and Erica Nelson, who got lost in Denali National Park in Alaska. For six days they rationed out the food they'd taken for what they thought would be a half day's hike; the newspaper article listed this food—candy bars and the like. The search and rescue of the girls, the article said, cost $118,000 and involved one hundred people. Turns out that the SAR (Search and Rescue) teams were able to locate the girls only after one of them got a cell phone signal and was able to call her mother and then text rescuers as a method to save her phone battery. So it was cell phone savvy that saved them, not outdoor survival skills. There's a photo of the girls smiling, one of the fathers hugging them. Nelson, 23, said, "I actually didn't think we were lost." Flantz said she thought they would "bring some more food" *next time.* I huffed to myself, reading the article: how did they get themselves in such a fix? What silliness. And not even embarrassed. Now I was wondering what it would cost to rescue me. I began to wonder what the newspaper headline would say. "Kentuckian stranded on Rogue Island without paddle: Claimed he wanted to defeat river." "Author of canoe narratives no match for Rogue; your taxes pay for rescue." And so on. It would be one thing if this was the real Rogue, upriver, the wild and scenic Rogue. I was dealing with tailwaters, next to a freaking highway. My predicament could not even be dignified with the word *wilderness*.

I'd been there over an hour. Sat on the bow of the boat with my head in my hands. Noted a piece of driftwood, a stick that looked fairly strong. Next to it was some fishing line, fouled and knotted. A couple of pieces of bark, like planks. Willows grew in clusters on green, bendable stalks. I began to break them off and wrap them around a piece of the bark, attempting to affix it to the end of the stick. I wound the willows in "X" patterns around the bark and tied them off at the end. Quite a few of them broke as I attempted to tie square knots in them, but I got enough of them on there for the bark to hold. I reinforced this with the fishing line. The other piece of bark took more time. I felt that I needed a double-bladed paddle in this rough water. With just enough steering capability to float to the end of the island and to the right bank somewhere, I could get out and off this damn river. After I finished the paddle, I tested it in the current, standing on the island. One piece of bark stayed in place, the other, the left side, rotated as I pulled it through the water, more or less useless. I reinforced it with more fishing line, pulling so hard that the line sliced into the side of each hand. Worse than a paper cut, these lacerations would fester for a few days, a healthy reminder.

It took me a few minutes to work up the nerve to get into the boat, but once I was in, I had no choice but to go forward. The drift boat was still ahead, in the same place. I took two strokes. On the second stroke, the left hand paddle blade fell off. Frantically, I paddled to the bank and pulled the kayak back to the point of the island, stumbling in the current, strong here even next to the island. While I worked on the paddle blade, I heard a scraping sound. The yellow boat had dislodged itself and was drifting, ever so slowly, alongside the island. I stomped after it, dove into the water and caught it. More shivering now, more chattering of teeth. Desperation bordering on despair. I figured out that the willows worked better if I peeled off the fibrous, pliable bark and used that to bind the paddle blades instead of the limb. After a few minutes more of clumsy wrapping around the driftwood, I got into the boat again. I used my paddle strokes sparingly. Let the Rogue have its way with me. Past the point of the island I went. Past the drift boat, conscious of the stares, but not looking back, focusing on the right bank, a place where the bank sloped down and offered water that might hold me there and not push me forward to doom. The paddle held together, worked just as well as the store-bought one. I wish now that I'd kept it.

I clambered onto shore and pulled the boat up onto the muddy bank. I was on a dirt track somewhere, a mile or so from the road, so happy to be off the island, happier and more grateful than I remembered being in a long time. The other problems that remained—getting back to my car and asking permission to trespass and returning for the boat—seemed miniscule next to the fact that I was off the island, off the river. No more swimming in the Rogue. My mouth was dry. I needed a beer. Or two.

I walked up the road to a house and knocked on a door. No answer, but a truck was coming down the road. It was a couple, and they did not seem surprised to see a wet person approaching them. Without speaking to me, the woman went inside to unload the groceries, and the guy listened to my story. I'd lost my paddle but was able to get off the river, stupid of me to be on it in the first place in my boat.

"What kind of paddle did you have?" he asked.

I told him.

"I could lend you one," he said.

"No, I've had enough of that river," I said. "That river's too much for me and my boat. All I wanted was to get permission to drive down here and load my boat."

"Go get your boat, and I'll drive you back to your car," he said. He was smiling. "My name's Leonard."

The place on the river was his brother's. He and his wife had moved here to take care of it; they were from Montana.

I bragged on Montana. He said the winters were a bit much for him. It was milder here.

I told him how foolish I'd felt on the island, stranded. "There were ten year olds rocketing past me on rafts," I said.

"Kids," he said. "They're fearless." The river, he said, can put you in serious trouble. Not long ago, a trio had arrived at his house near dark, wet and shivering, a girl in hysterics. Their raft had been punctured, and they swam the same section I had. "This weekend," he said, "somebody will have to call the rescue squad a couple of times."

I tried to give him a five-dollar bill at the parking lot, something my dad had taught me to do, but he would not take it. I loaded the boat, as if it had been a normal day, and went off in search of my paddle. At the campground where I'd planned to take out, there was nobody standing around with a paddle, no paddle floating at the ramp for me to pick up. I drove further downriver, to Shady Cove, and asked a fisherman if he'd seen a paddle. He said he saw another drift boat pick up a paddle, and they took out back at the campground.

"The river was a bit too much for me," I said.

"I've seen other people in those hard boats on the river. One guy said he was going to paddle to the end of the river. I wondered if he knew about the dams downriver. Good luck finding your paddle," he said.

I drove back to the campground, and there at the ramp was a drift boat with two fishermen. One of them got out of the boat with a paddle.

"Did you find that paddle in the water?" I asked.

"Is it yours?"

I said that it was.

"You got here just in time," he said. "We found it swimming down the river."

"That's what I was doing, too."

These guys didn't laugh. They didn't find me amusing, and I think they were on the verge of saying "finder's keepers." Anyway, it was a minor miracle that I recovered the paddle and a major miracle that I got it back so quickly.

My next stop was the Shady Cove grocery store, where I purchased bratwurst, buns, and two tall cans of beer. On the drive back to the campsite, I was still so exhilarated by my immersion/self-rescue that I called my wife, who was working and didn't answer, then Mel, a friend since childhood who had played professional tennis and now coached the Murray State team. After listening to the story, his only comment: "See, that's why I don't do that shit."

I cooked the brats over a driftwood fire and sipped the beers under the shade of "Doug" fir. My back hurt a little, and I had a scrape on my arm and scratches here and there, but other than that, I was uninjured. There would be no newspaper articles, no expensive rescues. I contemplated staying another day to rent a raft and make it down the Rogue without a crisis, like the fearless Oregon ten year olds, but the more I thought about it, the stupider it seemed. Tomorrow I would continue south and west out of Oregon and camp somewhere on solid ground to wax poetic beneath the redwood canopy. Somehow, in spite of Oregon John's sarcasm and my hostile attitude, it seemed the logical thing to do.

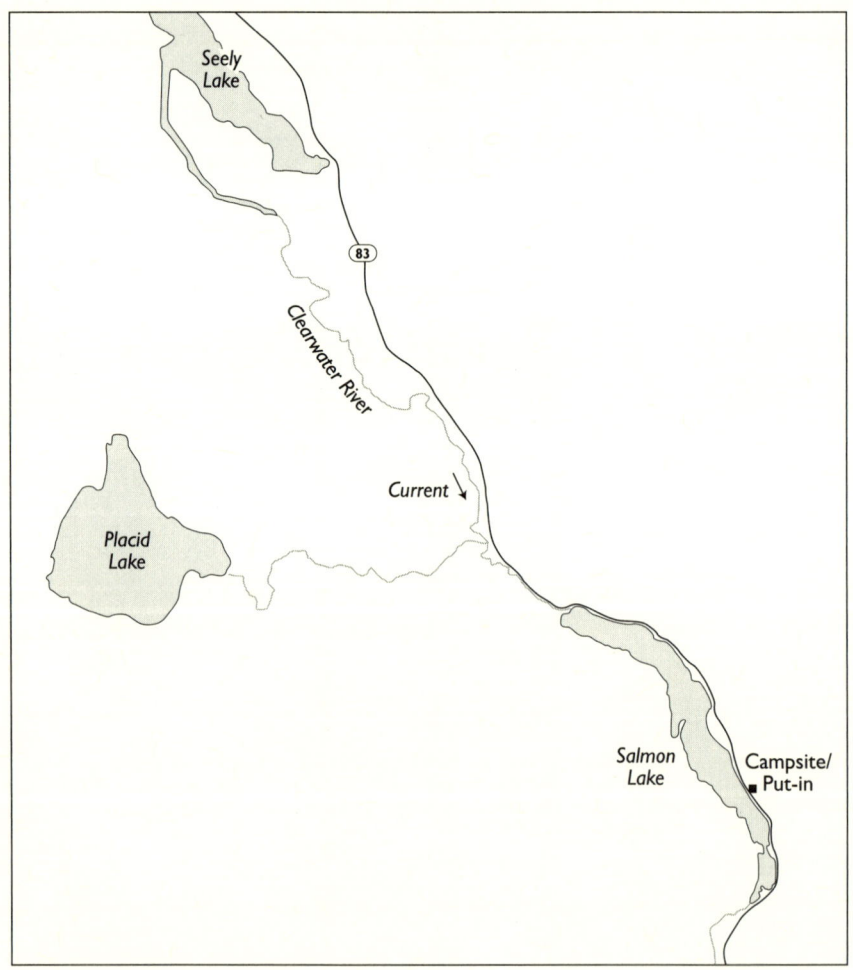

Salmon Lake/Clearwater River

CHAPTER 7

AESTHETIC CONVERGENCE: THE CLEARWATER AND THE DESCHUTES

After you take a long trip, people always want to know the best place you visited, the one you remember the most. That's almost always a complicated question for me, particularly on multiweek trips, which usually yield a bunch of favorites. After this loop around America, having sampled dozens of rivers and lakes, I had no trouble naming my favorite place, and I always used the name of the lake to identify it, not the name of the river, though as names, each is mundane in its literalness. Nobody I'd talked to had even heard of the place, much less visited there to confirm my taste in campground/lake/river, and because of this, I mentioned Salmon Lake less and less as the winner of my summer 2008 landscape beauty contest; I felt as if I were publicizing and demystifying a place and an experience that I should keep pure and unadulterated, though as I found out later, this is pure foolishness. Not only is Salmon Lake the next lake over from novelist Norman Maclean's hangout, Seeley Lake, but the Big Blackfoot, of which Salmon Lake's feeder river, the Clearwater, is a tributary, was featured in the movie version of Maclean's novel, *A River Runs through It*. That Salmon Lake was in the proximity of celebrity landscapes made it no less affecting, and in the following narratives I attempt to articulate and justify my choice for "Best Looking"—Salmon Lake/the Clearwater River—and "First Runner-up"—Lake Billy Chinook/the Deschutes River.

Salmon Lake/Clearwater River

In Missouri Headwaters State Park, where I set up camp, rain and wind kept up off and on all day. I positioned my tent up near a thicket of willows and parked my car nearby in an attempt to create a shelter from the wind, which seemed to swoop in from all directions. The temperature was in the low forties at mid afternoon, and I worried that the hoodie, light windbreaker, and two pairs of long pants that I'd brought might not keep me cozy through the night. I drove into Three Forks just to see how much a room would be. At the only hotel I could find, the owner/desk clerk was taking his sweet time with the customer who'd gotten there just before me.

"I thought I'd have to take out a loan at the bank to fill up my truck," said the tourist, there for the trout fishing.

The owner said he'd gone to the local bank the other day and one of the tellers said to him, "You seem like you're having a bad day, friend."

"I said to her, 'I am not your friend, nor will I ever be your friend, bitch. No friend of mine would work in a bank. You people giving out all these loans have caused the mess we're all in today.'"

He went on like that for a while, recalling the dialogue, and even if he didn't lace the original conversation with obscenities, he was laying them on now. He hated bankers and those who worked in banks. I started to edge out when he began a disquisition on state taxes and state government, but the fisherman got his bill paid, and it was my turn. I had tried to calculate the charge for a room by watching the exchange of money and thought the guy got forty dollars back from a hundred.

"What's a room?" I asked. "Sixty?"

"Eighty," he said without looking me in the eye. Hardly a welcome tone. I had on a long-sleeved purple T-shirt that my sister had given me: "Save What's Left," it said. I glanced around the lobby, at the tables and chairs and stuffed animal heads mounted on the walls. The conversation would be lively, and I'd gotten myself sufficiently worked up in the car listening to Dennis Miller, the snide and smartly funny guy from *Saturday Night Live* who had (apparently) converted to conservatism. The ideological war was raging full force now that Obama had gotten the Democratic Party nomination for president. Was it safe to talk to anyone these days about politics? Only those you agreed with, it seemed.

I thanked the contentious hotel owner and went back to my cold and rainy twenty-dollar campsite, where Lewis and Clark had traveled upriver until they reached this, the source of America's third longest river, the first white people to see it. It was the summer of 1805. They'd lost only one of their crew of thirty, to a ruptured appendix, and the other men had suffered from a variety of ailments such as boils, frostbite, and insect bites, but they had survived the passage through the Great Plains and the intimidating tactics of the mighty Sioux, and they had wintered among the friendly Mandans in North Dakota, where they took on an important traveling companion, guide, and interpreter: Sacagawea, a Shoshone who had lived near the headwaters as a child. At the age of twelve, she had been captured by the Hidatsa, who were attacking her tribe. Having been bought and sold a few times, she met Lewis and Clark's Corps of Discovery near present-day Bismarck, North Dakota. She was sixteen at the time, and her husband was French fur trapper Toussaint Charbonneau, with whom she had a child that Clark called "Little Pomp." Pomp was the infant who traveled with her on this historic voyage.

When the rain let up, I hiked down the Madison River and up onto Fort Rock. From there I looked across the Gallatin to the cliff top—Lewis Rock—where Meriwether Lewis looked down at the Missouri River headwaters for the first time, one of the major goals of the Corps of Discovery. Rust-colored moss splotched the gray rounded rocks on the ledges and soft aster petals bloomed purple. Bugle-shaped flowers drooped indigo in pockets of rich soil among thin knee-high grass. All around me were snow-covered peaks: the Tobacco Root and Madison ranges to the south, the Big Belt Mountains to the northeast. Below, the rivers ran strong and wide, flashing over rocks and stumps between brushy banks and marsh. In Yellowstone Park to the southeast, it had snowed six inches the night before, according to a couple I met at Confluence Point, where the Jefferson meets the Madison. I strode through the drizzle in my black poncho, snapping photos of sharp-billed black-and-white mergansers; western tanagers, flashy yellow birds with red heads; and a deer browsing a thicketed island across the Madison. I was the only tent camper at the small campground, and there was only one RV there, the host, who had not even poked his head outside. This was a place freighted with history, and I loved having it to myself. Even in the dreary weather, it was full of color, movement, and energy.

The wind rushed against the nylon tent and I lay there, wearing all the clothes I owned, listening to AM radio. A tornado had swept through a Boy Scout campground in Little Sioux, Iowa. Four were killed, many others injured. The young scouts were cited for their bravery and first-aid skills. The boys who were killed had taken shelter in a building the tornado had crushed. Holed up in my tent, the sleeping bag over my head, I stayed comfortable as the temperature dipped into the mid thirties. Lewis and Clark had written quite often about the insects that plagued them, mosquitoes in particular. One good thing about this weather: no bug bites.

After discovering the headwaters, the Corps of Discovery headed southwest up the Jefferson on the way across the Bitterroots and on to the Pacific coast. I headed north, down the Missouri, on Highway 287. I turned off on a gravel road that led to Toston Dam, the first of many impeding the burgeoning flow of the great river. I planned to put in above the Toston, which created the first slack-water lake on the Missouri, and paddle upstream until I could paddle no more. The weather remained overcast, windy, and cool. To the west the Tobacco Root Mountains were capped with snow and shrouded in ragged gray clouds. The dam itself was unassuming, a low structure protruding ten or fifteen feet above the surface of the river, seventy-five yards across. The boat ramp was closed, a sign in front of it warning that river conditions were hazardous and that there were no "boat barriers" installed at the dam. In other words, there was no way to stop boaters who got swept toward the dam, especially those without motors. I considered chancing it for a ways, but given that the river was not pooled behind the dam, since it seemed to be running wide open through the six floodgates, well, there would be no sense of discovery as I plowed upriver in the easy water near the bank. I drove on, considering Canyon Ferry Lake downriver. The camping was so bleak there, the lake combed over with harsh winds, the campground nearly treeless, the highway within plain sight, that I kept going northwest on Highway 12, past Helena, then up 141 into the Garnet Range.

Like Meriwether Lewis, who was given to spiraling moods, I was disappointed. I hadn't explored a transitional zone since Lake Shafer, back in Indiana, and the two rivers I had considered since—the Cheyenne and the lower Missouri—had been unsuitable for my purposes. So I kept driving. I wandered in the general direction of the Columbia River above the Grand Coulee Dam near the Canadian border. Where I would camp next, I wasn't sure, though I had a vague notion of going north to vast Flathead Lake and paddling up the North Fork of the Flathead River. Along the way I passed through some of the most striking country I'd ever seen, where the Big Blackfoot River and its tributaries flowed crookedly across grassy meadows. It helped also that the sun flashed between the clouds now and then, the sky playing out its drama of shadow and light across the deep green of the fields. I stopped at a campground beside the Big Blackfoot and watched it rage, the biggest river waves I'd seen so far on the trip. A sign warning of grizzly bears made me think of how far I'd come from New Jersey, where a ranger had warned me about black bear sightings. Nobody was camped here. It would be just me and the bears. Little did I know that I'd stumbled onto a movie set. I wasn't aware of it at the time, but the Big Blackfoot is not only one of the most popular fly-fishing rivers in Montana, it is the river in which Brad Pitt, at the direction of Robert Redford, hooked a monster trout and got pulled down the river, swimming in water over his head to retrieve it.

I turned north on Highway 83, still heading toward Flathead Lake when I saw a state park near the road on a ridge thick with lodgepole pines. As I drove, I glimpsed between the narrow tree trunks blue Salmon Lake glittering diamonds of light, then the Clearwater River, which fed the lake, wending its way through the willows. I turned around. It was the best decision I'd made in over a week, stopping at this lake I did not know existed, and the important thing that I needed to remind myself was that it was a decision that did not require much contemplation. I had come across Salmon Lake by chance, and I knew nothing about it other than the brief glimpse and the emotions that arose from this glimpse.

The two rangers—tall, bearded Chris and a young woman, Jessie—arrived in a pickup as I was sliding my camping fee into the box. They said that indeed the lake had stocked salmon. Also trout and bass. Someone had "introduced" northern pike, said the ranger, and they were threatening the salmon. Teeth marks had been found on harvested salmon. I asked the whereabouts of the dam, and Chris informed me, without condescension, that I was camped beside a glacial lake, that the Clearwater River flowed into the mile-long slack water (created by the scouring of a glacier over the millennia) and then flowed right back out on its way south to the Blackfoot. This was Clearwater Junction, where a couple of well-known explorers had preceded me. It seemed that without really trying, I was shadowing Lewis and his men, who came through the area in 1805, on the way back east, Clark having taken a group, including Sacajawea, on the same route they'd come west on.

I set up camp under the ten-story-tall pines. Shafts of sunshine splashed patches of warm light on the brown pine needles. So pure and bright here, the sun lifted my mood after what had seemed like weeks of rain, including the catastrophic flood back in Mauston, Wisconsin, and a bleak rainy campsite on the outskirts of Bowman, North Dakota. Here the pines whistled in gusts of cool wind. The constancy of wind was something I'd forgotten about the West, at least on a visceral level, and here it was, a reminder that, warmed by the midday sun, a thirty-mile-an-hour gust of wind gave you a hint of possibilities, that you needed fleece, boots, hat, and firewood when you stayed outdoors here in the summer. I gathered some damp firewood, which was plentiful, but I could not resist the automated wood dispenser. Four dollars' worth of quarters activated the bin, which rolled forward a bundle of immaculately split pine logs and opened a door at the bottom for harvesting. There was something appealing about a vending machine that dispensed wood instead of stale crackers and Twinkies.

Buck, a civil engineer from Mississippi, was camp host, living in an RV with his wife, Lacy, a nurse. They had been taking these kinds of jobs, he said, for the past two summers. An experienced bass fisherman, Buck was anxious to catch one of the northern pike. He'd heard they were full of fight. There was a canoe trail, he said, on Seeley Lake, about fifteen miles away. You could put in, float a few miles on the Clearwater, and take out just a short walk from the put-in. He said he needed to get his canoe in the water soon. It had been too long between canoe runs.

In the town of Seeley Lake, I stopped at the visitor center, a converted cedar-shingled barn. On the first floor a museum was partitioned off into horse stalls with themes: logging, hunting, and fishing outfitters; Native Americans; agriculture; and so on. Rudimentary farming implements hung in one stall, animal pelts in another. A bear reared up, forepaws stretched out toward me. Stepping into the third stall, I realized that I'd stumbled upon the land of Norman Maclean, whose book is the *Moby-Dick* of trout fishing. His family had a cabin on Seeley Lake, about ten minutes away. I was so surprised and pleased with myself for having stumbled on Maclean country that I browsed over the museum mementoes quickly and rushed to get outdoors.

The town of Seeley Lake, in Swan Valley, was pure blue highway. It had a good drive-in hamburger/milk-shake joint, a small Laundromat, a non-supermarket grocery, and a bar or two. Something about the town, the campsite, and the surprise of the literary landscape put me in a buoyant mood. Driving up the campsite road, I hailed Buck, who was standing on top of his RV taking something that his wife handed up from the ladder.

"I think you should be on the canoe trail today," I said.

He looked down at me. "What's that?"

I repeated myself. No reaction from him, not even a nod.

"Are you from Oxford?" I asked, following up on the fact that he'd told me he was from Mississippi.

"I've spent some time there," he said in what seemed to me a summative, dismissive manner.

I nodded and drove on, wondering what I'd done to annoy him. I was into the third week of camping alone, silent except for the sounds inside my head and the noise I made chewing, drinking, walking, and organizing my way across the country. Maybe he hadn't recognized me in my sunglasses and cap. Maybe, just maybe, I had become the scraggly loner who people wanted to avoid. Was I frightening them? Upon further reflection, I guessed that I'd hollered at him at a time when he didn't want to be bothered, completing some chore that perhaps his wife had asked him to do. Maybe they were quarreling. I thought that I might have violated what seemed to be emerging as "western campground" etiquette. You said hi to people, and you were friendly when approached, but you didn't poke into their business. You left them alone generally, and you were sensitive to their need to be left alone.

For supper I combined a can of turnip greens and a can of hominy, and then added a chopped onion. Not too bad, but not a do-again. A dark squirrel seemed to think it a delicacy. He circled my camp for an hour, making kamikaze runs under the picnic table where I ate. I gathered an arsenal of pine cones and began tossing them at him as he approached, hiding behind tree trunks to dodge my artillery. Except for the camp host and one other group of campers, who had an RV with a rather loud generator, it was just me and the dark squirrel.

Next morning I carried my boat and gear three hundred yards across a slope of dry slippery pine needles to the put-in, a low bank of big yellow flowers. My breath steamed into the air, the temperature in the high thirties, the sun a couple of hours away from striking the lake. Far above, in the sun, a jet trailed a thin white line, silent against the blue sky. The lake was calm, and as the sun rose above the steep slopes, it cast crystalline reflections of cloud and sky upon the water. Up ahead three miles, hidden within thickets of willow, the Clearwater awaited. Three openings gradually emerged as I approached, trying to paddle without making a noise in the deep silence of the windless morning. The far left, my choice, wound around, mazelike, twenty yards at its widest, what we in the South would call a crick, a brook in New England. I kept thinking I'd see a grizzly as I rounded each bend, and I was able to paddle a mile or so into the maze before the current began to shove the bow back and forth at tight bends, and I let it push me back to a little dock, where an upturned canoe lay on the bank beneath ponderosa pines, no cabins, no beaten path to the dock, a notable absence of "NO TRESPASSING" signage. There I ate a peanut butter sandwich and listened to a variety of birdsong.

There was something pure and unfiltered about this river/lake that all dammed reservoirs lacked. On the return I found a split in the river and came out on the lake at a different spot than I went in, nearer the highway. It was the best transitional zone so far, a liminal experience, nothing freakish or dramatic, tranquil and pure and understated. I was completely alone on this lake, within the deepest, most profound silence of the trip. Even though a two-lane highway was within my sight for much of the voyage, I heard nothing but a few birds, the remote shush of a fish breaking the surface, and the small splashes my paddle made as I submerged it, pulled it through the water, and withdrew it from the lake, creating silent whirlpools as my wake. For one morning the place was mine. There were pleasing contrasts: the dark green of the mountains against the evanescent wavering quality of the blue water, the wide glassy lake and the tumbling boiling current winding through the maze of willows. It was hard to sum it all up, but I think that the symmetry in the lake's reflection of the sky and the intensity and sharpness of light made this such a memorable paddle.

My notion of the sublime, so well-represented here by rugged and pure Salmon Lake, has its roots in eighteenth-century Romantic poets such as Wordsworth, and because these criteria—which tend to include precipices, waterfalls, snowy mountains, raging waters—were articulated some three-hundred-odd years ago, it's worth questioning their contemporary validity. Jonathan Raban's essay "The Curse of the Sublime" (collected in his *Driving Home: An American Journey*) brings into question many of the criteria that we use to set landscapes apart and to protect them from development and commercial use. Raban noted, first of all, the elitism of America's popularizer of the sublime, John Muir, who, starting in the late nineteenth century, was one of the earliest developers and promoters of our National Park system. Early on he championed Yosemite and wrote much about its majesty and nobility and isolation, language that still survives among contemporary nature writers. Raban sums up Muir's thesis: "in the craggy aristocracy of the peaks and woods, we commune with majesty and nobility, and thereby rouse something noble and majestic in ourselves" (353). Then he documents Muir's cultural and racial elitism, noting his disdain for the Native Americans who had lived in the vicinity and his championing of the aristocratic values he saw when he visited the South. Raban asserts that Muir's legacy is one that "divides the nature tourists and conservationists from the majority of people who live and work in landscapes cherished by visitors for their grandeur" (354). This divide, this elitism, Raban concludes, makes environmentalism, an otherwise worthwhile cause, ineffective. If we justify our love and protection of wild places "in terms of the ennobling spiritual benefits of roadless hiking, and the snobbish taste for natural beauty of the urban leisure class," we risk widening the divide between preachy tree huggers and people trying to make a living off the land. Raban's suspicions of the "messianic" environmentalist and the link between nature and spirituality begun by the British Romantics, continued by American Transcendentalists such as Thoreau and Emerson, and popularized by Muir, is clear.

That the notions of Wordsworth, Coleridge, and Shelley—Englishmen like Raban—led to a sort of elitism among those who subscribe to it may vary with whoever is beholding the land. I was willing to share the sublime with others, though it was true, I didn't want to see it overrun with visitors like me (an easterner starving for the depopulated purity of the West), and I'd rather the ridges rising from the lake were not clear-cut to provide jobs for Montanans. I very much appreciated that the lake was formed by a glacier, and not by a dam, even though dam building creates jobs. But I was still puzzling over all of this. What constituted the sublime was part of what drove me on this quest for liminal zones, and articulating what physical dimensions, what history, what confluences of time and place produced it was complex and in no way an exact formula influenced by any one school of thinking. Having been on Salmon Lake, I might conclude of the sublime: "I know it when I see it, and this is it."

That evening Buck visited. I told him how far I'd been on the trip and that this was my favorite campsite so far, for the clean bathrooms, the private showers with the hotel-worthy water pressure, the plentiful hot water, and, best of all, the automated wood dispenser. I told him I'd paid twenty-eight dollars in Ohio on the Mohican, and the showers were slimy, though there had been a swimming pool. The most he'd paid for camping was thirty-five dollars a night on the coast of northern California at a site that didn't even have flush toilets. He said it was worth it. If you go farther west, he said, you must camp somewhere on that northern coast. This was one piece of advice I heeded, one of many "must-see's" that I paid attention to. The conversation veered to dangerous animals. He and Lacy went for evening drives up into the mountains, looking for wildlife. They spotted a mountain lion down a slope fifty yards from the road, licking its fur "like a house cat," said Buck. Something got its attention, and it transformed itself, moving forward, shoulders hunched, its belly low to the ground. Buck, an experienced hunter, said this made him nervous, this stalking carnivore, even though he and Lacy were sitting in a vehicle. "If they decide to come after you," he said, "they'll catch you." He said the cat weighed about 150 pounds, but the ones that hunters killed were as big as 250. He'd tried to get a photo of it, but every time he looked through the viewfinder, he couldn't frame it because it blended in so well with its surroundings. He hadn't seen grizzlies in the area, but they were around; black bears were common, though he'd seen none at the campsite in the few weeks he'd worked there.

That night, my last on Salmon Lake, I piled crackly pine onto my fire and fell asleep in my camp chair. In the morning I noted quite a few burn holes in my ground tarp, my tent, even a couple on the sleeve of my windbreaker. Not smart. I'd been thinking of Maclean's second book, *Young Men and Fire,* about the Mann Gulch fire a few miles east of here, near Helena, where twelve smoke jumpers died in a blowup, a sudden conflagration of wind and fire. Intoxicated by the fast burning pine and its companionable sound effects, I'd built my own fire beyond the limits of common sense. Such are the dangers of trying to enhance the sublime come nightfall.

I stayed two nights at Salmon Lake and might have stayed longer and explored the other lakes in the chain that the Clearwater fed, but I could not shake my fear that familiarity with the place might subtract from its quality. It wasn't perfect, by any means, and many of the things I liked—such as the practicalities and novelties of the campground, the conversations with a kindred spirit—were rather mundane and obvious. The place was so pure and unrippling in its beauty that first morning that I was afraid that more exploration might lead to a hidden carcass, a trash heap, or a disillusioning human with information that might taint my brief impression. There's a time to linger and seek knowledge and a time to move on, to let a memory be. Such is the prerogative of the lone traveler. A reasonable companion might have persuaded me to stay longer.

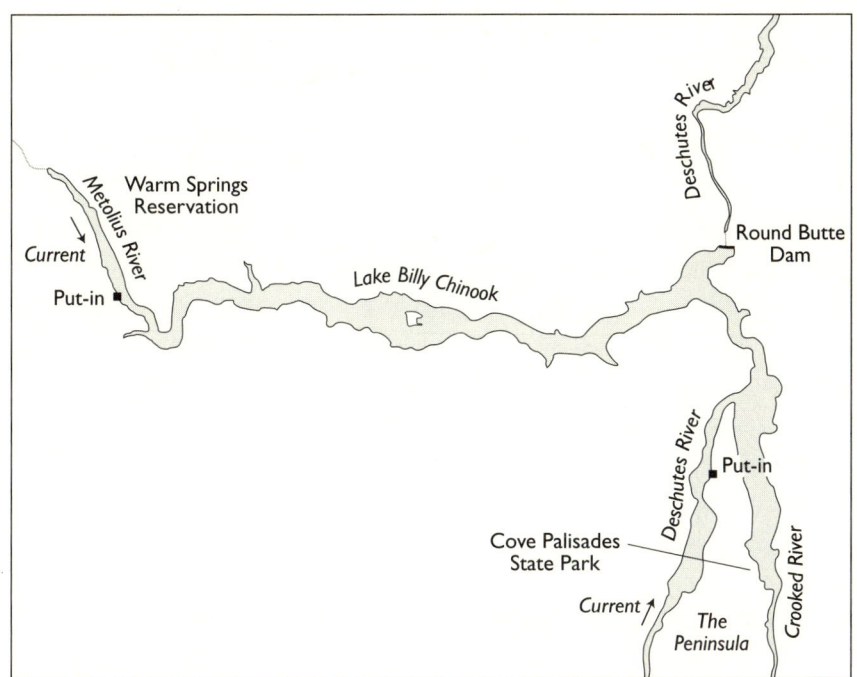

Lake Billy Chinook/Deschutes River

Lake Billy Chinook/The Deschutes River

After Salmon Lake I thought it unlikely I'd find a place where the planets aligned to give me perfect weather, a near-perfect campground, and a pure and natural liminal zone, its impact on me still lingering, a bit of a mystery, like the mazes of willows that led to the refreshening of the Clearwater. It took me over a week of meandering in a generally northwest direction, to stumble upon the runner-up in the Liminal Zone Landscape Sweepstakes.

As I headed away from the Columbia on Highway 97, I sliced through the searing high desert of Oregon, where the wind scoured the dusty landscape. I felt the desolation internally, as if I were swallowing up the emptiness all around me. After a few hours of extreme highway solitude, rarely even a passing car, I entered Madras, where at the visitor center I inquired about an intriguing lake I'd seen on the map at the border of Warm Springs Indian Reservation. Lake Billy Chinook is fed by three rivers: the Deschutes, the Crooked, and the Metolius. An unlikely trio, it would seem, from their names. The Metolius's name was derived from a Native American term meaning "spawning salmon." Hudson Bay Company fur traders—Frenchies once again—named the Deschutes, meaning river of chutes or falls. The Crooked? Self-explanatory.

The young woman at the visitor center loaded me down with maps, magazines, and brochures. "Could I paddle my kayak up the Deschutes?" I asked, peering at her around the stack of glossy paper.

"Absolutely," she said. "You can go up the river quite a ways." She knew because she'd done it a while back, in a canoe. This was the first upstream paddler I'd met, at least the first who admitted to it, and she bubbled over with western buoyancy, counteracting the funk I'd contracted in the high desert drive.

Cove Palisades State Park seemed to me a rather grandiose name, given the landscape I'd just passed through to get here, so I was skeptical as I drove there from Madras. Once I came upon it, I let my guard down. It was as if the crust of earth had been cut open and folded back to reveal a secret: a shimmering dark-blue lake fed by three rivers that twist through canyon walls dotted with juniper, silver-green sagebrush, grasses, moss, and desert flowers, crowned with towers of tan basalt—the palisades. A nineteenth-century homesteader had named a swimming hole on the Crooked River "the Cove," and the place where two rivers contributed to the flow of the major river—the Deschutes—had long been a gathering place for Native Americans. A Wasco scout, Billy Chinook, led explorer, surveyor and mapmaker John C. Fremont to this place in the 1840s.

In 1964 the confluence of the three rivers was made into a reservoir with the completion of Round Butte Dam. The lake was named after the aforementioned Wasco scout. I wondered what Billy Chinook himself would think of the dam and the deep water, compared to what the place had looked like before the damming. The water wouldn't be nearly as deep as it is now, four hundred feet at the base of the dam. From the canyon rim, the rivers would flow like small ribbons, farther below, shimmering for brief periods each day when the sun found its way to the bottom. Engineers thought it would take over a year to fill the reservoir, but because of heavy snow and rain in 1964, it took only a week and a half. Now the controlled flow of the three rivers provides power for the citizens of Portland, Oregon. The Native Americans of Warm Springs Reservation, whose lands border the Metolius, help to manage the dam.

My assignment? To paddle from the artificial "arms" of the rivers as far as they let me. I camped at the fringes of the first campground I came to, at the lip of gorge. My site was grassy and cool beneath the pale-green leaves of Russian olive trees, an irrigated field a few yards away, beyond a barbed wire fence. To the west, fifty-odd miles away, loomed Mount Jefferson, part of the Cascades, a conical peak almost completely covered in snow, its gray rock jutting through the white cloak like the mountain's scaffolding. It glowed light pink as the sun sank below the horizon just north of the peak. My campground, on the rim above the Crooked River, was full, and I did not see another tent camper among all of the RVs. The campground below, near the Deschutes, had tent campers, and it was next to a popular trail that skirted the perimeter of the canyon rim, but there was no view of Mount Jefferson. For that I would not mind being the odd man out under the stars, the only guy who didn't have the pumps and hoses and canopies and plug-ins. Next to me was one of a couple of sets of camp hosts, from Texas, who set off twice a day in their golf cart, destination unknown, and the rest of the time stayed inside. On the other side, a fifth wheeler family had set up its canopy and accoutrements on the side away from my little site, so that when I looked that way all I saw was the rendering of a Kodiak bear snarling within a marketer's conception of an Alaskan landscape.

One couple sat in camp chairs beside a trailer that looked like a rolling A-frame chalet for dwarves, just big enough for a short person to stand up in. They had a pair of kayaks turned upside down on the ground, so I asked them which river arm they'd paddled and how far. The woman, Joy, said she'd paddled around on the lake. They lived in Portland. When I told her husband, Jim, that I was paddling up dammed rivers, that I was looking for the places where the current started up again, he became animated. He worked for Bonneville Power Company and began to describe the intricacies of buying and selling power generated from dams. He said that an aluminum smelter near Billy Chinook got a government discount on power and signed a lease for it that extended over decades. Instead of using all of the power, the aluminum company bought more than it needed and sold its surplus on the market. After a time they had made enough money to shut down, and the workers were sent home to a lucrative retirement.

I was on the Deschutes arm by six the next morning, in a hooded sweatshirt and long pants. The temperature was in the low forties. Thirty-odd miles upriver from Round Butte Dam, the river lay still but narrow between canyon walls. After a mile of paddling, I passed under a small highway bridge and continued another two miles. A rock outcropping in the middle of the river told me that I was close. Current swirled around this little gray rock. Up ahead fifty yards the river poured over a chain of rocks that extended across the width of the channel, about thirty feet wide. I found an eddy on the left bank and got onto shore to contemplate the transitional zone of the Deschutes. Alders and pines grew along the flat bank downriver and leaned out over the current, a change from the nearly treeless canyon walls. Breaking its silence during the three-mile paddle, the river made a washing sound as it poured over the rocks. There was a chute about ten feet wide between pointed rocks where a downstream paddler could pass, but the current, coming around a sharp bend, would be too resistant for me to fight my way upstream. Upriver from where I sat, the sun had begun to illuminate slivers of water, but I was within the shadows, and the glassy surface reflected the sunlit red basalt and sky above in sharp relief. As I paddled downstream, the warmth crept nearer and nearer, spreading down each canyon wall, at one point casting its glow upon a doe and two fawns who paused to consider me below as they browsed the grasses, headed the opposite way. It had been the shortest journey from reservoir to river that I had taken, and the trip back was quicker than I wanted it to be.

Under the spell of the mirrorlike Deschutes, I decided to stay two more nights at Cove Palisades, the view of Mount Jefferson and the prospect of two more river arms enough to persuade me to risk a letdown from the excellent impression I'd gotten on that first day. This was the first dammed lake I'd been on since Indiana, and I had to admit that, as lakes go, I loved it. I'd begun to realize that the lakes themselves were important elements in my judging the overall quality of a transitional zone quest. My template had been of going from the sullied and manufactured and polluted still water to the revelatory mystery and naturalness of moving water. Now I was beginning to see that even a dammed lake such as Billy Chinook could be beautiful, and the beauty and drama of a western lake deep in a gorge took nothing away from the discovery of the river that fed it.

Kenny, the clerk at the camp office, had a strong opinion about what I should do during my extended stay. He thought I should drive the forty-five minutes to the Metolius arm, which was more like a "real river" than either the Crooked, the nearest, or the Deschutes.

"I mean," he said, black hair sticking out from under the bill of his cap, "it's not like a canyon filled with water." What he didn't realize was that a canyon filled with water seemed remarkable to an easterner who was used to the vast, spread-out TVA lakes where a bluff forty feet tall became an object of desire for boaters and cliff jumpers for miles and miles, a monument to landscape relief.

Kenny was so enthusiastic about the Metolius and sketched me such explicit directions that I felt compelled to try it, even though the mouth of the Crooked lay only a few minutes below us.

"Have you kayaked the Metolius?" I asked.

He had kayaked once. In Sweden. "It rained so hard," he said, "that I've never kayaked again."

Kenny's coworker Chris came to relieve him. He was a little older. As soon as his government check came in, he said, he was going to buy a pontoon to fish the lake. He was most interested in catching trout, he said, because he "liked to eat trout." I admired the simplicity of this principle but wondered why he didn't like to eat salmon from the lake. Chris specialized in bull trout, which didn't sound appetizing, though they were also known by the more whimsical "Dolly Varden."

After the subterranean mystery of the Deschutes, I modulated my expectations of the Metolius. Beware, I thought, of falling under the influence of other people's enthusiasm, no matter how well meaning or expert they are. Eating dust on a gravel road for dozens of miles, I arrived at a forest service campground and tried to sneak down the gravel that crunched under my tires as I rolled past tent campers, most of them still sleeping, one woman drowsily rekindling the previous night's campfire. It was a thickly wooded campground, every site sunken in dark shadows. I put in at a ramp on a bay and was immediately confused about which way was up. Upriver, that is. The bay was at a big bend, and there was no current. After paddling about a mile in the wrong direction, I scrutinized the map and determined that I needed to paddle east, back along the shore that was lined with cabins on one side, sagebrush and cut timber on the other—reservation land. On the developed side, I began to see a PRIVATE PROPERTY/NO TRESPASSING sign about every fifty yards. And just when I was on the verge of entering what looked like a promising transitional zone, complete with nesting bald eagles, I was confronted with more signs on the reservation side. One cedar sign said, "NO TRESPASSING," and below it in smaller letters, "CLOSED TO ALL FISHING," then "350 YARDS TO CABLE CROSSING." A smaller metal sign on the same post alarmed me: "NO TRESPASSING," it reiterated, "CLOSED AREA. VIOLATORS SUBJECT TO FEDERAL OR TRIBAL LAWS." As if that weren't enough, where the river narrowed beside a boggy island stood yet another sign, its post standing in water, driven into the bed of the river: it repeated the "closed area" message. Now, it seemed to me, someone wanted to make it clear that not only were the banks off limits, but also the river itself. I didn't think that anyone, native or not, could own a waterway, but I decided not to risk it. The phrase "Tribal Laws" dissuaded me more than "Federal," and the closed area message made me wonder if there were some kind of wildlife sanctuary beyond or if I would be disturbing something fragile with my paddle strokes. Randy Russell and I had unwittingly paddled into a carp sanctuary upriver from Nashville, on the Cumberland River/Old Hickory Lake, and we were scolded by the ranger there, who made us carry our canoe a half mile, away from the carp haven, when we left. Here it looked like I might get more than a scolding, and I certainly did not want to be subject to tribal laws. Three tribes made up the population of the reservation—the Wasco, the Warm Springs (formerly Walla Walla), and the Paiute. Given the historical information below, from the reservation's website, it's understandable that they're protective of their land:

> Under the treaty [of 1855], the Warm Springs and Wasco tribes relinquished approximately ten million acres of land, but reserved the Warm Springs Reservation for their exclusive use. The tribes also kept their rights to harvest fish, game and other foods off the reservation in their usual and accustomed places.

There is no mention of the Wasco scout Billy Chinook on the historical section of the website, only the observation that by the mid–nineteenth century "waves of immigrants" disrupted "the old way of life for the Indian bands in Oregon."

On the return trip, a large mammal ogled me from a long pier extending from the bank. Except for a buffalo in Idaho and a large-scale human or two, it was the biggest thing I'd seen the entire trip. It easily would outweigh the deer of the previous day. This, my friends, was a river otter, who waited until I was within twenty feet to squat and defecate on the dock, rolling off into the river just as I got my camera out. At first I thought it was a seal. It must have weighed seventy-five pounds.

As I approached the boat ramp, a child's voice screeched across the bay. "Mommy, look at the man in the little boat!"

The mother's softer voice explained, I supposed, what she thought I was doing. A dog worked its way down the steep bank to swim. It was odd, after observing so much, to find oneself a part of someone else's tableau. That evening I hiked along the rim of the canyon on an ancient trail—the Tam-a-Lau ("place of big rocks on the ground")—crows swooping across the chasm below me, their shapes deep black against the glimmering green river far below. A boat and skier drew a sharp white line across the water. Also below me was the island, a plateau of basalt lava flow that had been closed to visitors since 1997. One of the few areas that hadn't been grazed by cattle, these isolated two hundred acres retain a rare ecosystem whose grasses, herbs and shrubs resembled the presettlement West.

In my immediate future was the near drowning in the Rogue River (see chapter 6), the giant redwoods and Sequoyah trees, and the California coast. Though recognized by many as life-altering places/events, they would not surpass for me the aesthetic quality of Salmon Lake and the Deschutes River, remarkable for the ease with which I reached the liminal zones, the drama of light and shadow, the contrast between lake and river, and the general absence of human engineering in each place. The more I contemplated the places, the more I thought about them after I'd left, the more convinced I became that I would find nothing else like them.

CHAPTER 8

RECONSIDERING THE LIMINAL: THE DOLORES, THE CONEJOS, AND A FRACTIOUS CAMPGROUND IN FOLSOM, CALIFORNIA

These three places—Folsom Lake in California and the Dolores and Conejos rivers in Colorado—got me rethinking the concept of the liminal. By the time I got to the town of Folsom (yes, that Folsom, famous for its prison), which was roughly the halfway point in my loop, I'd found what I considered the most beautiful liminal zones I'd seen (see chapter 7), and I'd altered the trip according to whim, not only discovering remarkable and secluded places that I didn't know existed, such as Big Lagoon (see chapter 9) and Idaho's St. Joe River, but also realizing that my definition of a liminal zone as this specific place between a lake and a river might be overly limited.

Folsom State Park

Strange messages came from my radio as I headed east on Highway 36 through the Trinity National Forest and across the Mad, the Trinity, and the Van Duzen, the kind of rivers you see in coffee-table books full of large glossy photos. Far below me they ran, invisible in their private gorges. This particular AM station seemed to have chosen me, its signal alone penetrating the canyons and forests with enough clarity for me to hear bluegrass music, the banjos and high-pitched harmonies at odds with the landscape and the sophisticated culture of northern California.

"We are receiving reports of a major drug bust involving the FBI and local law enforcement," said the DJ. "We are informing local citizens of roadblocks in the area as we receive reliable information. This is a bust of major proportions. We want you to be aware of this situation, should you see unusual activity by law enforcement."

I was on my way south from the Humboldt County area to the Sacramento airport to pick up my wife, Julie, and spend a few days in wine country. I was looking forward to eating and drinking well, to driving up and down the California Coast with Julie, and to stopping when we felt like it, without an itinerary or a "to do" list. I was looking forward to her company, but I had to drive through this hell to get to her. I had one more night of camping—and I didn't know where yet—before picking her up the following morning.

Every five minutes or so the banjos on the radio would stop, and the DJ would break in with more information, sometimes admitting that the people calling in might not be absolutely reliable. There were checkpoints where cops were armed with assault rifles. There were reports of a police convoy. And, by the way, he would interject every so often, don't try to get to Shasta Lake because of the fires. They are difficult to contain because it is so hot and dry, and they are in places so remote that firefighters can't get there to fight them. Then he'd go back to the drug bust and tell us about a report he'd gotten that the FBI had gone into someone's home to search. If they come to your house, he advised, you do not have to answer any questions, but "do not physically obstruct them in any way." Then back to songs about moonshine and trains and tragic love.

The *Times-Standard,* Eureka's newspaper, reported the next day that, in raids across Humboldt and Mendocino Counties, the feds had seized 25 to 60 million dollars' worth of marijuana plants, plus a vehicle, computers, cash, and thirty weapons. Personnel from the IRS, the Postal Service, the Drug Enforcement Agency, and the California National Guard Counter Drug Task Force had gathered in Fortuna to coordinate the sting against an association of growers that had purchased two thousand acres of mountainous land "for large scale commercial marijuana production." To clarify the intent of the operation, Special Agent in Charge Charlene Thornton said, "This is not a medical marijuana operation or a group of people growing for personal use. The targets of our investigation are reaping huge profits while contributing to the crime and violence oppressing communities across the state."

As the airwaves crackled with this Humboldt County drama, I passed into Trinity County and got a lot more interested in the fires. Smoke lay heavy in the valleys, at first a whitish haze like a really bad August day in the Tennessee Valley, then a brownish atmosphere with no breaks in it. It was so thick that I couldn't see more than several yards ahead on the road. I met no vehicles other than a couple of fire trucks freighting water. I raised my windows and turned on the air conditioner, then drove faster than I should have up and down curvy roads, thinking that I could drive out of it, that it would ease up once I got back to civilization, represented by Red Bluff and Interstate 5. At one point I got out on the side of the road and experienced a heat such that I've never felt, a stifling, penetrating atmosphere that reached down and seized my throat and lungs. Brittle grasslike tinder covered the hillsides. I never saw flames, but the fires couldn't have been far away. Men worked along the side of the road in helmets and uniforms, but considering the smoke and the scale of the multiple fires—more than eight hundred at this point, started by the "dry lightning" strikes of the previous weekend—the human effort seemed miniscule. I wasn't quite panicking as I drove through the smoke toward Red Bluff, but I did drive with some urgency and purpose, never having seen such a hellish landscape on such a grand scale. I'd driven almost a hundred miles through constant smoke.

I decided against my original plan to visit Lake Oroville, which was in the thicker smoke up north, near Paradise and the western slope of the Sierra Nevada. Around Sacramento the smoke thinned out a bit, and I pulled into Folsom Lake State Park, which smoldered in the low nineties. I drove past campsites perched on dusty little hillocks, most of them in full sun or latticed by scraggly juniper or spruce. At the bottom of the campground loop, in a grove of cedars, I chose a site at the edge of a dry grassy gulch with a path that led to a thicker corridor of trees. I unloaded my kayak and much of my gear, set up my tent, and drove to town to wash the bugs and dust off my car and to vacuum up the worst of the pine needles so that my wife would not be embarrassed to ride around with me in uppity wine country. It was bad enough, I guessed, to have the yellow boat crowning my humble Subaru.

When I returned to camp, a husky guy emerged from the rear of a van at the site next to me and greeted me with this observation: "I guess that's a good theory. They're more likely to rip you off in town than they are here." He was referring to the fact that I'd left my boat and gear unattended in camp.

I walked over. He grew up in Folsom, he said, and he was back, "on leave" from his job, to take care of his mother. Because he'd gotten into a disagreement with her neighbor, he decided to camp here for a night to avoid potential confrontations. He wore shorts, hiking boots, and a t-shirt. There was no tent.

"This is not my life," he said, gesturing at his belongings in the van, as if to say, "I don't live out of my van."

He was a location manager for a prime-time TV reality show, and inside his van he had film-editing equipment. He told me about his work and about the people and situations he had to manage to prepare for a show, and he confided that he aspired to make his own films. A tree standing a few yards away caught his attention. "I remember that tree from my childhood," he said. "I'm going to shoot it before it gets away." He told me he would shoot it sideways and then run the film through backwards or some such jargon I didn't quite get.

I told him I was from Tennessee. "Oh, really?" he said. "I couldn't tell from your accent." Grinning. For some reason my drawl often elicited derision from Californians. I remember my cousins, who lived in Beverly Hills, making a big deal out of it during my visits.

I told him what I was doing on the trip, where I was headed, and he had much advice. "Put your video camera in a plastic bag, and you can shoot underwater if you want," he said.

"Stay out of Napa Valley. It will just depress you. Go to the coast and then north, away from the fires." He told me that I must stop at the Hearst Mansion, "that or head east to Lake Tahoe. You do not want to stay in Napa Valley," he warned.

As I was receiving all of this advice and mentally drawing red lines through it as if to delete each item, an argument flared up among a group that included three men and a woman at a site just down the road from us. The woman stood apart from the men, at the rear of a dusty maroon van, a silver ladder on its door. She seemed to be the focus of the dispute in front of the van, and one man raged at another, the third man trying to separate the two suitors. There was much stumbling, much swearing, much pushing and pointing of fingers. The location manager kept delivering travel advice to me as this drama threatened to escalate from pushing and shouting to something involving punches and kicks or perhaps even weapons.

"There's a reality show over there," I said.

"That's not my reality," said the location manager flatly. "Doesn't interest me." He kept talking about what I should see next in California and how to avoid the fires. The shouting escalated. One man pushed another with force enough to knock him down.

"That's starting to worry me," I said, unable to ignore it.

The location manager leaned toward me, his face inches from mine. "Don't even look over there," he whispered fiercely. "That is not our concern. There's good people, and there's bad people, and there's white trash." I started to point out that the disputers were Latino. Perhaps he said "white trash" as a generic term that he could be sure I'd understand, though I really didn't understand where white trash stood between or beyond good and bad. Maybe he was calling me white trash. What he said next precluded my request for clarification.

"If you go over there and try to do something, they will kill you," he said. "Even if they see you put a phone to your ear, they will shoot your ass." He said this firmly, with conviction. After a few more minutes of pushing, the men seemed to tire, and the wronged one, the jealous one, sat at the picnic table, his head between his knees, staring into the dust.

I excused myself to fix dinner, backing away from the location manager, who disappeared into the back of his van for more editing. Canned Heat began to blare from the open back doors. Meanwhile, at the site on the other side of me, just a few feet away, camped a beautiful couple, a tall dark man who resembled Tyrone Power, and his wife, a slim, brittle-looking blonde who chain-smoked. There weren't saying much to each other. The man left in a battered white pickup and returned with three kids in the back, all of them sporting long, dark wet hair, tall and slim with beautiful skin. A boy, about fifteen, used "f——g" every third word and directed much of his wilting derision at his younger brother, a chubbier version of himself. The older boy was long-limbed with straight black hair cascading onto his shoulders, and he had a cool, slouching prowl when he moved, which wasn't often. The girl was maybe seventeen. She sulked, occasionally screeching out a "motherf——r" when one of her brothers called her a "ho." It was the unhappiest family I'd ever seen, their dysfunction all the more poignant because they were so beautiful. Unlike the outburst of drunken passion across the way, to which the family seemed oblivious, these tirades of derision, insult, and protest seemed ritualistic. The blonde wife tried to reason with the daughter, and though Tyrone Power posed in exasperated stances, arms folded, and requested in a low voice of authority certain behavior from the kids, the parents never raised their voices, and nothing they said was heeded. The children ruled. I tried not to stare. The family barely looked my way, but the location manager was on his way over again.

I had a tall beer now, and he was deeper into whatever he'd been drinking.

"I put that music on to calm them down," he said, "and it worked. See?" We glanced at the quartet, subdued by Canned Heat and Neil Young. They had moved apart from each other, lulled into separate reveries. "Music calms the savage beast [sic]," he said.

He was up to telling me his life story. As a youngster, he got a job at a copper smelter in Ely, Nevada. He came back to California and worked as a contractor in the Los Angeles area, and that's how he got the location manager job, almost by accident, showing up at the right time to substitute for a no-show and staying on and learning the trade on the job. Now he was back home, taking care of his mother. His older brother and sister took no responsibility for her, and somebody had to. She had Alzheimer's. He talked a bit more about his job and about his leave of absence. He gave me a business card.

The romantic campsite seemed to have arranged a truce, and all of them were getting drunker and drunker. Marshmallows were roasted among the unhappy family, and the boy told of a gym teacher who said he looked like a girl. "I wanted to hit the stupid motherf——r," he said to his parents. "Im'a f——g kick your ass," he kept saying to his little brother.

The location manager seemed to understand that I wanted to rest, so he let me be and disappeared into the back of his van. I got into my tent, reluctant to turn on the radio for fear that I'd miss a noise that I should heed. At this agitated campground, in a shallow doze, I had a sacred flying dream. In it I flew low over the crowns of giant oak trees and dilapidated barns, wearing a thick canvas suit with broad wings that held the wind and let me hover and dip and dive in the air like a crow. It was the kind of dream I hated to wake from, and I wondered if it came from a longing for home. Sometime that night, the location manager left. That morning, the mother and daughter were up early. It was the daughter's birthday. She was supposed to go to school, it seemed, but she protested tearfully that she preferred not to. She and her mother left in the car. When she came back, two hours later, she opened an envelope from Grandma with a card in it.

"How much did you get?" asked the older boy.

She told him. "Happy birthday," he said sadly, his first utterance that lacked an obscenity.

Across the way the romantic campsite was quiet. One of them slept under the picnic table, another wrapped in a blanket in the dust near the fire pit. The van's sliding doors were open on both sides, revealing two slumbering forms. Bottles lay about.

I had time to kill before picking up Julie at the airport that evening. I went for a run, trusting that the dysfunctional family was too self-absorbed to "rip you off," as the location manager put it. I ran down a paved trail to the dam, where workers were digging postholes for a new gate across an access road. Beyond stretched Folsom Lake, which backed up the Sacramento River here, a dull gray body of water spread out under the haze of smoke and humidity—not one boat disturbing its surface. I took a dirt path through some dry brush and squat trees. Ahead stood a cat; he crept into the weeds as I approached and raised his head, ears pricked, when I ran past him. He was about the size of a cocker spaniel, and I was thinking "bobcat," as we stared at one another; he was about fifteen yards away, with tiny white chevrons on each ear. I continued on a loop that took me back to the workers.

"Are there bobcats around here?" I asked.

"Did you see that cat?" one of them said. "We saw him, too. He had a long tail on him. I think it was a mountain lion. That's what the ranger said. We called him."

"I'm glad he didn't jump me," I said. "That would be a helluva way to go."

"I would have taken your picture," said the guy.

"Thanks," I said. I ran back to the campsite, took a shower, and gathered my gear. The daughter was lamenting the fact that she had not brought any clothes to change into after she'd showered. The brothers offered clothes. She rejected their offers with disdain. One called her a "ho" again. Hysterics ensued, and while the father, cooking pancakes, brooded at the periphery, the mother raised her voice for the first time. I can't remember what she said, only that it was a tired, desperate sound that earned her a few moments of acquiescent silence.

Where would they spend the rest of the day? What would bring them peace if a camping trip aroused such animosity? I packed up and headed toward the Roseville Public Library, my refuge from the Folsom camping experience. The mountain lion sighting haunted me, its matter-of-fact appearance in my path, its crouching in the patch of weeds between a dam and a four-lane highway and a campground. In its own way, it seemed as desperate as the campers there.

On the way to Roseville, I drove the four-lane highway past upscale shopping centers. I was in the right lane next to a truck that maintained the speed limit, and we traveled side-by-side for a short stretch. As if in the Daytona 500, a gleaming black Mustang veered from the left lane to within inches of my bumper. He eased off, then zoomed up again, and I've never seen a vehicle that close to mine at fifty-five miles an hour. He was sending me a message loud and clear to speed up and clear the way so that he could motor ahead of us and be the first to the stoplight. He found a gap between me and the truck, and passed, missing each of us by inches, but he did not make the green light. I pulled up beside him. He glared at me, a well-groomed man, dark hair, his insolence complementing his fine features. Something about the guy made me hold his stare and lower my window. He lowered his passenger window.

"Is there a problem?" I asked.

He waited a beat, considering. He shook his head. "No," he said, somewhat petulantly, I thought, as if I was in the wrong, as if he hadn't just terrorized me with the bumper maneuver.

"Good," I said and raised my window as the light turned green. He zoomed off at a high speed, his final statement. Never have I done such a thing. And although my intent was to hold him accountable for aggressive, unreasonable driving, I knew the moment that I lowered my window that my action would be deemed the aggressive one, the act that could escalate into a violent confrontation, as it often did in this land of speedy traffic and high stress. I'd let the lack of sleep and the influence of my camping neighbors bleed over into my trip.

While waiting for the library to open, I ate breakfast at a restaurant run by a Korean man. I was the only customer. The breakfast—bacon, eggs, toast—wasn't spectacular, but it soothed me. He said that he didn't do much business at breakfast and that times were hard, jobs scarce. No one passed by on the sidewalk as I ate. I spent the afternoon in the library and in the city park, where some people seemed to be residing. The descent into the Sacramento Valley had deflated the western buoyancy I'd noticed for the past two weeks and landed me squarely in an urban purgatory. In some ways Folsom and Roseville seemed a liminal zone that I had stumbled upon, a place neither urban or rural, neither civilized or wilderness, but containing elements of each extreme, where I had no expectations, as I would, say, at Crater Lake or the redwood forest. In particular the flying dream and the wild cat—the forlorn predator— seemed liminal emblems at this dusty campsite, these transcendent, unexpected experiences so poignant and fragile within the matrix of anger, heat, and desperation. I needed some good wine. I needed a break from my solitary wanderings and to fall under the civilizing influence of my wife. And after that I needed to discover a more uplifting liminal zone and linger under its spell.

The Dolores River

Common sense would tell you not to paddle a kayak after dark (and before light) on a lake you don't know, but since when did common sense know how to have fun? A wind kicked up, just a breeze, but enough to add some quirks and visual trickery to the landscape. I broke through waves in absolute isolation on this dark moonless morning. Across the lake I angled in a straight line toward the opposite shore, two miles away. How deep was the water? Careful, careful. An unidentifiable object began to rise from land that jutted out from that far shore. The object blocked the eastern horizon, where the lid of night was slowly, slowly sliding back across the sky, the coming daylight faint, an insinuation of dawn. Behind me, all was dark, like a cave from which I'd emerged. Hypnotized by the waves and wind and the dark water, I began to dwell on the name—Dolores, sorrow, dolor—and the voyage took on a macabre tone. Sorrow in the dark, sorrow for all the bad things that happen to people, no fault of their own, no matter that they did silly things like kayak alone in the dark (the result of setting my alarm to the wrong time zone). We're sorry, didn't anyone tell him to stay off the water before daybreak? You can run over things you might not see. They can tip you over. In the dark.

Edward Abbey, writer and cranky environmentalist, steered a raft of passengers down the Dolores one of the last times before the river's "damnation," the earth-fill dam built to irrigate sorghum and alfalfa and to supply the factories of what he called "Cortez, the Shithead Capitol of Dipstick County, Colorado." Abbey was not one to rationalize the altering of nature, particularly when it was a river he loved, and this one, he noted that morning, was lovely: "Snow melt from the San Juan Mountains creates a river in flood, and the cold waters slide past the willows, hiss upon the gravel bars, thunder and roar among the rocks in a foaming chaos of exaltation" (231). Abbey knew the Tellico River, where my quest for transitional zones began. Someone sent him a newspaper clipping that described the barns, the houses, the fertile land, and the Native American burial sites that would be obliterated under the permanent flood called Tellico Lake. He admired a young man who chained himself to a rock below the high-water line of a coming reservoir on the Stanislaus River in California, if only to halt the stilling of the waters for a time, before he could be found and removed. What does it mean to love a river? It helps to know it well, I think, to spend time with it, to learn its moods in different seasons, to boat it and to swim it, maybe even drink a little of it. You love rivers you grow up near, where things happen to you, good and bad, and the river becomes a backdrop that makes what happened more vivid, perhaps more beautiful in memory than in the moment. The Dolores I'd only known a short time, but the day before, as I sat in the town's public library and watched the current shoot past just a few feet away, I began to develop a bit of a crush. I told the librarian that this was, without doubt, the best public library I'd ever visited, and I'd seen a lot of them. She was glad to hear it and told me how the money was raised, about the donations of building materials and labor, and about the pitch she used to raise contributions. The wood floor was donated, and someone had built the fireplace and an enormous stone chimney for free. But the best part was the location; the panoramic windows faced the river so that you could sit and read or daydream while scanning the ever-changing patterns of moving water.

 At the Dolores Chamber of Commerce, a little white house with a low chain link fence around it and American and Colorado flags flying in the front yard, I asked the two women where I might be able to rent an inflatable kayak for a run down the Dolores. "You can't here. You have to go to Durango for that," they said. I considered aloud putting my own boat in and floating a couple of miles.

 "The water is high," said one woman. "I wouldn't do it. Just last week, the rescue squad had to pull a man who had been tubing from the rocks below the bridge. It's dangerous now."

 "I wouldn't do it," said the other woman. "Especially not alone."

 Though I was itching to get on the river, I told them I'd paddle the lake instead. This pleased them. They were so happy I thought they might kiss me. Abbey's ghost shook his head in disgust, his arms folded, his beard long and majestic. I was sorely tempted to go ahead anyway. Throughout the day, I explored the river on foot, a jog up and down it. I sat in the library and stared at it. It was fast and strewn with rocks, but most of the passages looked simple—stay left, stay right, shoot the middle—at least it looked that way from the bank. If I fell out, I reasoned, I wouldn't be swimming like I did in the tailwaters of the mighty Rogue. I'd stand up and get back in the boat. Yeah, or crack your skull or tailbone, said another voice in my head. Down the river, near the bridge, a scattering of boulders would require more intricate maneuvering, the waves higher and deeper. More doubt crept in, along with memories of the Mohican and the Rogue and the way my boat misbehaved. The way I had behaved. Skirt or no skirt, I decided against a downstream journey. Abbey turned his head and spat on a rock. But the great one, he was never alone on a river, it seemed. In the desert, yes, but not on a river. I had no friends here, and all the people I met told me to stay off this river. I'd capsized and swam twice on this trip. The third one could be it. I could hear what the Chamber of Commerce women might say to a reporter: "We told him not to do it, and he said he wouldn't. He didn't seem right in the head, though. It seemed like he was listening to someone else when we explained how high the river was."

If Dolores the town had benefited economically from the dam, it hadn't stuck. Earlier I'd stopped at a store that advertised outfitter services. Idle gas pumps sat outside, and the store itself gave off a forlorn, neglected air. I had to speak up to rouse the owner, who seemed to have been dozing in the back. He didn't outfit anymore, and he'd long since quit fixing people up with fishing guides. Durango, up the road, had put him out of business, he said.

"Don't know why they bother hiring those guys in Durango," he added. "They don't know anything about the lake."

A brewery nearby seemed out of business or perhaps it had gone dormant during what should have been its busiest time of year. That afternoon I walked into a bar that had a bicycle, a motorcycle, and a couple of pickups parked on the street next to it. Inside two guys played eight ball, and a couple of old guys whom I joined at the bar seemed like they'd been there forever. Fixtures. On the television, outdoor bowling. Beer signs blinked. The billiard balls clack-clacked, soft cursing ensued. Elk and deer stared down from the walls, and pheasants were caught frozen in mid-flight near the Budweiser clock. The bartender listened to one old guy's troubles and seemed surprised when I didn't order a second beer. I imagined she was used to people lingering and either making a pass or trying to think of something witty. Or staring, tipping, staring some more, for she was easy to stare at, a kind of visual oasis among the usual bar stuff. At the gas station, I bought a tall beer to take back to camp, and the clerks inside seemed unaccustomed to making change. A Native American with two kids was trying to pay for candy, fishing worms, and gas. He shook his head and rolled his eyes at me as the clerk tried to get the change right. The other clerk was struggling with the outside payers. My lone beer grew warm, and one of the kids kept looking from the can up to my face and back to the can, as if wondering if I could drink the whole thing.

Despite its troubles, I loved the town of Dolores. There was a restaurant called the Naked Moose. There was no McDonald's. And it had the best public library of any small town I'd visited. If I became homeless, I would move to Dolores and spend my days in the library, my nights on the periphery of the campground overlooking the lake. Look out, Dolores.

After such extensive scouting of the lake and the river, I had a problem. I'd already seen the liminal zone from the highway bridge, so the point of my quest was somewhat moot. To transform things, I left the next morning an hour before first light. The unidentified object I was headed for delineated itself against the faintly gray sky: a chain-link fence, ten feet high, was surrounding a peninsula. I could not see what was being fenced in or what I was being fenced out from. I hoped the daylight would reveal whatever the fence protected us from.

At the bridge, current swirled lazily around the piers, Dolores rousing herself. I veered to the right bank, where the river began to turn due east toward her namesake town, then Stoner, and then Mount Wilson, where the river originates. I paddled into a marsh and picked my way through the narrows between half-submerged sagebrush and fallen trees, high water once again, which was commonplace this summer. A beaver brought his tail down, and it made that glugging sound of something being plunged hard underwater, like the open palm splash we all made in the bathtub. Another one. And another. Ahead a great blue heron perched on a limb that reached out from the bank, his long-beaked head resting on his breast. Not yet arisen, this slumbering heron, too early for fishing. Without moving my arms, I glided closer to him in the still water of the slough, the wind blocked by the sheltering ridge above. Something woke him, maybe a smell or the color of my boat, and he raised his head, extended his neck, and ever so slowly lifted himself into the air with heavy wings, creating his own wind as he lumbered up the slough, awake now, dammit, but withholding that wonderful hoarse squawk. I disturbed him twice more before having to turn around, my passage blocked by trees and brush. I would never tire of watching him lift off into that majestic flight, but some sort of nature etiquette told me that enough was enough. Back on the river proper, I paddled another fifty yards before the Dolores's bottom came up, rocks across the channel under the next little highway bridge, a mountain stream again, such a sudden transformation from the lake behind me, like the change from night to day, from tunnel to open ground.

Back at the ramp, I pulled out and sat on a flat rock. Bud Light cans decorated the shoreline. Nearby lay a swimsuit top, brand name "No Boundaries." I could understand her shedding the top, even to fling it onto the ground, but to abandon it completely? No boundaries indeed. As the morning advanced to a civilized hour, a man with a white kayak thirty feet long pulled down to the end of the ramp. By the time I got up to go over and talk to him, he had launched the boat and gone somewhere up or down the river, I couldn't tell, because he had disappeared. Another man scanned the ground with a metal detector, headphones snug, communications down.

The Fourth of July loomed. I headed east into the San Juan Mountains to escape all the fuss and the fireworks.

Platoro Lake/The Conejos River

On the wall of the Starbucks in Durango, the Humane Society had posted a photo of Bullet, a shepherd mix posed in profile, ears laid back, a noble dog in need of a home. Bullet's sad and stately portrait haunted me that day and a few days afterward. It made me realize how much I missed my own dog, Jasper, who had been dead for a year. In his prime he would love a trip like this. He would have made many friends, and these people would talk to me because of him. In a sense, having Jasper along on trips like this helped to validate me in the eyes of most people. Alone, crossing the country, no real destination? No dog? Not even a cat? A bit strange. Bullet, I realized, could never replace Jasper, but he had that knowing look that rare dogs possess, the ones who seem to size you up and think beyond their species. Without Bullet, without sampling the wonderfully named Animus River, which roared through town carrying rafts full of wet, laughing people, I drove east out of Durango, toward the isolation of the San Juans and what I hoped would be a transformative liminal zone.

On my road atlas, Platoro Reservoir, on the southeast tip of the San Juan Range, near the headwaters of the Rio Grande, seemed a quiet place to spend the Fourth. The Conejos River fed the reservoir, which was a few miles

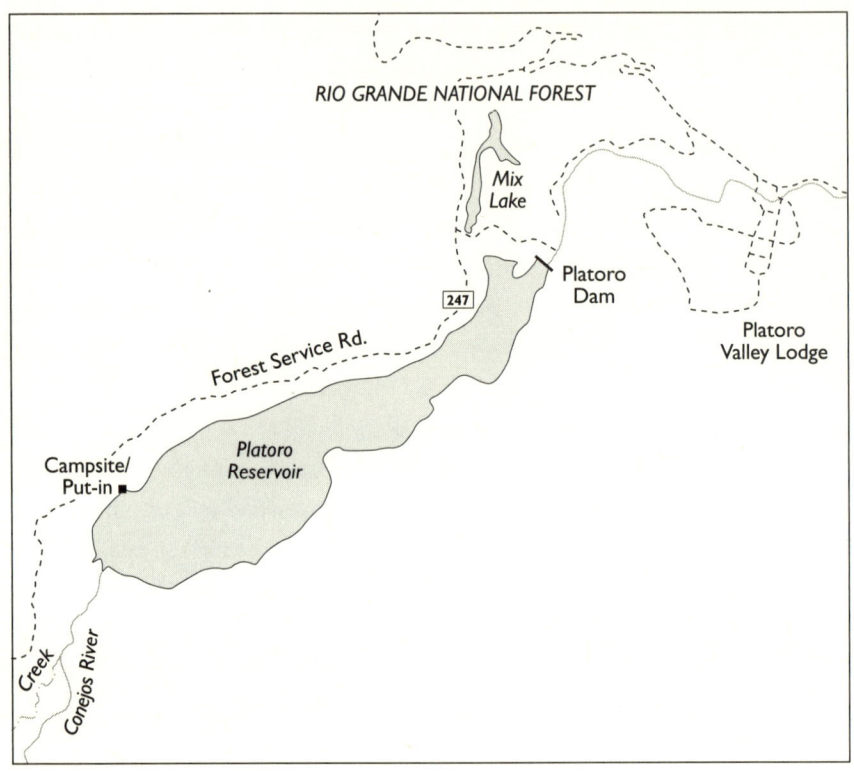

Platoro Lake/Conejos River

long, a half mile across at its widest. There were no paved roads that led to it, I could see that, but it looked like approaching it from the north on Highway 160 was about the same as approaching it from the south on Highway 17 in northern New Mexico. I climbed over Wolf Creek Pass, where the snow lay heavy at 10,856 feet. Descending from the pass toward the town of South Fork, I could not find a road that would take me through the Rio Grande National Forest to Platoro. So I kept driving on 160 until I stopped at Del Norte's visitor center, miraculously open on the Fourth. The woman, a transplant from the Texas Panhandle, first asked me what kind of car I was driving. A Subaru, she thought, would be acceptable, though I got the feeling that a jacked-up monster truck or a Land Rover might be better. Then she spread out *four* different maps, in addition to my Colorado map, and worked on finding a suitable route to the reservoir. One route, Blowout Pass, she ruled out because she thought it might still be blocked by snow. The other route approached Platoro from the west and looked to be about a fifty-mile drive. And it was about that distance, but it was the most arduous fifty miles I've ever driven. I climbed and plunged down a road that often seemed made of chunks of rocks scooped off mountainsides and plunked down, ungraded, a narrow, curvy incursion through the wilderness where washboard was almost a relief. At least washboard meant that someone else had passed through here. At about the halfway point, worried that I should have already arrived at Platoro, I stopped beside a creek where four camper trailers were parked in the woods.

"Just keep on going," said a man eating lunch. "I think you may have to take a left somewhere. It's been a while since I've been up there."

Up there. I had a quarter tank of gas, and my tires had worn pretty thin, so I worried about a blowout or a slow leak on these unforgiving roads. I babied the Subaru along about five to ten miles an hour. There was a place called Jasper, a small gathering of houses, where everyone I passed waved. "NO TRESPASSING. STAY ON ROAD" was repeated on signs during this stretch. A small gathering of kids and adults played wiffleball in a meadow. A man in cowboy boots, hat, and jeans drove a ball past the clearing into the brush. A woman named Delphine had carved her name into an aspen tree next to the road. Finally, a sign for Platoro popped up, and I turned left onto the steepest grade yet alongside a lake bluer than Salmon Lake, mountain peaks framing the far side of it, snow tucked in crevices that ran deep from the crowns of the peaks to the tree line halfway down. In a meadow that sloped from the road to the lake a half mile away browsed three elk, dark brown against the green of the low brush and grass. When I was seventy-five yards away, they looked up at me in my idling car and didn't budge. In a flat clearing full of bright-yellow dandelions, three modest camper trailers had parked in a circle, near the upper lake where the river and a creek came in. I parked nearest the road, about twenty yards from the nearest camper. I asked if I was crowding him.

He poked his head out a tent and said, "I couldn't even see you from here."

I set up camp and walked back up the road with my camera. The air's stark clarity sharpened the edges of everything, from the firs and pines to the snowy peaks to the dried blood spattered on the rocks next to the road, near an ATV whose seat had been torn off by the force of a crash. The elk posed, heads up, for a few moments before trotting off when a dust pluming sport-utility vehicle met me on the road. The driver, wearing a "U.S. Army" cap, lowered his window. Thin, with a round head and a sharp voice, he reminded me of the father on *That '70s Show*.

"Uh, we're checking for camera permits," he said.

I made a show of looking in my wallet. "What does one look like?" I asked.

"Like a twenty-dollar bill," he said. General laughter from the passengers followed, and he drove on, mission accomplished, big tires crunching gravel in small explosions. I walked another mile and started feeling weak and nauseous. I worried that eating the leftover beans and rice at lunch, a meal I had prepared the night before, had given me food poisoning. Had it spoiled? If I was sick, I'd picked a bad place for it. I sat on a rock and looked down at the lake, gathering my strength, and realizing after a time that my wooziness was probably brought on by the altitude—at ten thousand feet, Platoro is the highest artificial lake in Colorado. Also there was the fact that my main nourishment had been a "grande" coffee I had nursed between Durango and Platoro.

On the way back I scrutinized the ATV scenario. On the rock lay a fluorescent-green hanky with blots of blood on it. I was thinking head injury, fairly serious for the amount of blood staining the rocks. Somebody had abandoned the three-wheeler, which looked like it was undamaged except for the broken and detached plastic seat. Was this place too far off for the cops and the rescue squad to bother with?

Farther on, I met the SUV again, and this time the shotgun passenger, a meaty guy in a khaki bush hat like Teddy Roosevelt wore, said, "We're checking for beer permits."

"It's sad," I said, "but I don't have any beer." Silence from the SUV. I wondered if they were going to accuse me of treason for not drinking beer on the Fourth.

"Do you need a lift somewhere?" Laughter again.

I said that I didn't, that I liked walking.

They shook their heads at this last absurdity and drove on. Back at the campsite, near sunset, I headed down to the lake, a half mile hike. A group was fishing on the flat lowlands -where a creek entered the lake. Not far from the creek's mouth, the Conejos came in, bare of trees for the mile between the lake and the place where it seemed to emerge from the dark woods at the base of the mountain. This transitional zone was right in front of me, as it had been on the Dolores, within walking distance of my campsite. No mystery here.

A young guy in baggy, low-hanging shorts and a backwards cap was teaching three kids how to fish. He baited their hooks and coached their casting. They sat in camping chairs beside the mouth of the creek. He said he'd caught one rainbow trout.

"This place is beautiful, isn't it?" he asked, taking it all in after he cast his own line.

I agreed. It was something worth affirming, no matter how obvious. I asked them how they'd gotten here, and he said he really didn't know. He'd ridden along with his girlfriend's family from Albuquerque that day. They'd come in from Highway 17. He said he'd be out here at 4:00 in the morning to fish. "That's very early," I said. "I'll be out here at 6:00 in my boat."

That night I sat in the car with the windows open and listened to the Cubs/Cardinals game, coming in clear all the way from Chicago. Across the meadow a glowing Frisbee flew back and forth. Fires glowed from every site. One kid, gathering wood for his father's conflagration, said, "Dad we're going to melt the snow on the mountains." There were no fireworks. The stars, blinking fiercely, sufficed. I slept lightly, still wary about my stomach, not having eaten much. Sometime that night I sat upright, thinking I'd heard something opening the zipper of my rain fly. A big noise followed, the call of something large and wild. It sounded urgent. I shined the flashlight that I wore on my head out toward the small brook that trickled toward the lake, and two glowing eyes stared back at me. The thing seemed bigger than my car; it was bigger than my car. Mountain lion was my first thought, just surfacing from the fearful bleariness of sleep. Then it moved, an elk come to camp for a drink. I turned the light off and let him be, convinced, for no good reason, that elk were harmless. He moved heavily through the brush up toward a knob of pines. I was thinking that it would be good to have a dog with me, something with superior senses, able to alert me when things were moving through camp in a remote area such as this one, a designated wilderness area. Jasper, who used to sleep in the tent with me, far off in the corner, would have growled just loud enough to wake me long before the elk's big snort.

Next morning, I carried the kayak a half mile down a steep trail that descended thirty feet and then across a tundra to the point where the creek met the river. It was seven o'clock, and no one was up, not the fisherman who said he would be here at four and none of the other campers above on the shelf of land. My shoulders ached by the time I got to a flat place where I could put the boat onto the water. I was at the mouth of the Conejos in five minutes. It flowed fast and shallow into Platoro Lake, and it didn't tolerate my incursion for long. A mile or so upriver, it emerged from the forest, its headwaters somewhere in the peaks that loomed above. I let it carry me back to the lake, paddle across the gunnels, always the most contemplative time of my early morning voyages, so quiet it was, and easy, to float back on the same course, on the same river I'd just struggled against. Ease was a rare thing, more precious when it followed exertion. I paddled along the shore of the lake alongside trickling waterfalls. The lake was so clear I could see fish nuzzling the hull of the kayak and cruising farther below on the grassy bottom, oblivious to my passage in the world above them.

Back at the launching point, the young fisherman and his charges had reappeared. They may have thought me odd for having a boat and not fishing. In the lake were kokanee salmon, rainbow, brown, brook, and cutthroat trout, though mine was the only boat on the water that morning. Platoro rivaled Salmon Lake in Montana, with its clear water and postcardlike alpine scenery. That the Conejos was too small and swift to admit much exploration did little to diminish the Platoro experience.

The kids hadn't caught anything, but they hadn't complained. One of them had gotten into trouble for wading into the creek without permission. The water was very cold here, probably the coldest of the trip since it was at the highest elevation. Swimming was not high on my list of activities.

The young man, Mitchell, was a brick mason's apprentice in Albuquerque. Before that he had worked in an assisted-living facility, where he was shocked at the way the residents were treated. He would never let his grandmother end up in a place like that, he said. He'd also worked at some fast-food restaurants.

"I won't eat at any of those places," he said, "except for Sonic. Sonic is very clean."

He too had heard the elk rampaging throughout the campground the night before. He said someone had left bread outside, attracting the hungry beast. He and his girlfriend's family had admired the effect of the illuminated Frisbee toss the night before. This turned the talk to Roswell, New Mexico, and its reputation as a host to visiting aliens.

The "alien" at the museum there, said the girl, was nothing but "a dummy wrapped up with rags."

She said, "People in Roswell get mad if you question anything."

Mitchell got busy untangling lines and baiting hooks. I wished them luck and left, thinking what a good guy to be so patient with his girlfriend's siblings and in turn how calm and quiet the kids seemed sitting beside the clear creek. They were probably good kids anyway, Mitchell an exceptional young man, but you can't help but wonder about the positive influence of a place. How do we carry it with us back to the city, to our lives full of work and tedium and the small annoyances of school and work and neighborhood? How could you preserve such peace within?

I began my long portage alongside the creek to the campground. At the steep hill, I put the kayak on the ground and lowered the top of my head into the fast-flowing creek. I forced myself to stay there, kneeling in the gravel, my scalp tingling, then numbed. From this perspective, upside-down, the stream taking the sky's place, the sky below it, mountain peaks inverted, I tried to will the essence of the water's pure chill inside my head, the sound of it runneling between the gravel bars to the lake. When I raised up, the water trickling down my shoulders and back was so cold that I struggled to breathe, as if I were being born or reentering an atmosphere that I'd left behind for a timeless moment.

At noon I was on my way out, heading south to Highway 17. In Platoro, the town, which consisted of a store and a few cabins, a woman said there was a lot of washboard on the road I was taking, and we determined together that twenty miles an hour was the optimal speed over the humpy roads. When I told her I was from East Tennessee, she said, "That's a beautiful part of the country." It was, I agreed, but I couldn't imagine it standing up in comparison to a rarified place such as this. I wondered whether she had grown used to waking up here, if living here had made her long for a different landscape, say the vegetal lushness of the Appalachian Mountains.

More than any of my stops on this trip, Platoro Reservoir represented the sublime landscape, what Belden Lane would call "excessive beauty and grandeur," a conception that originated, he said, with those same Romantic poets that Raban made fun of. Because we have been acculturated to finding the sacred in the extraordinary, said Lane, we'd find it harder to accept it in ordinary landscapes, where beauty is more subtle than the Grand Canyon or the Pacific Coast, his examples. I understood his point about seeking the sacred in the ordinary, but, at the same time, Platoro, given its grandeur, struck me as a sacred place because of the difficulty of getting there and because of the interactions I had with the landscape, the wildlife, and the people I met. It probably helped, too, that I'd never heard of it two days before arriving, that I'd never seen a photograph or read a description of its grandeur.

Folsom State Park, the Dolores River, and Platoro Reservoir differed from other liminal zones because their effect upon me resulted at least partly from my manipulation of the experiences. At these places the transformative power of place was partially controlled by the way I approached and interpreted place and narrative. In all these places people played a greater role than place in the quality of the experience: the lost location manager and dysfunctional campers at Folsom, the endearing fishermen and obnoxious SUV riders at Platoro, and the Chamber of Commerce ladies and the ghost of Edward Abbey at Dolores. I'd been to more desolate places on this trip—the Humboldt River in the Nevada desert, for example—but these three places each held a unique desolation, and I was glad that I'd learned to be flexible about what I was looking for and what might emerge from the quest in unexpected ways.

(Opposite top) The transitional zone between Fontana Lake and the Nantahala River is one of the most dramatic I've ever seen. Here, shrouded in the fog rising from the cold mountain river water, fishermen toss lures below Greater Wesser Falls, just a bit upriver from the still waters of Fontana.

(Opposite bottom) On the 2008 trip from the east to the west coast, the Gauley River/Summersville Lake transitional zone was the first I encountered. It was beautiful . . . and deserted, as most paddlers headed for the violent waters below Summersville Dam.

(Top) Water released from Shafer Lake in northern Indiana shoots up in the air like a geyser below Norway Dam, not far from where I camped.

(Bottom) Human and nonhuman wildlife circulate among the merchandise at the White River Outpost of the Bass Pro Shop, upstream from Table Rock Lake and the James River arm.

(Opposite) This Hoosier on Shafer Lake strikes a permanent pose to attract boaters who need fuel for their engines.

(Opposite top) One of several practical craft negotiating the Rogue River, which had its way with me and my impractical hybrid kayak.

(Opposite bottom) Salmon Lake, formed by glaciers, not a dam, provided me with a near-perfect morning paddle on a chilly June morning.

(Bottom) The campground at Salmon Lake State Park features such amenities as an automatic wood dispenser, high-pressure showers, and a soft, fragrant carpet of ponderosa pine needles.

The Deschutes is one of three rivers feeding Lake Billy Chinook in Oregon.

Wind and calm create a distinct border on McPhee Reservoir near the Ute Mountain Indian Reservation in southwestern Colorado.

(Opposite top) A small group of campers joined me at southern Colorado's Platoro Reservoir, elevation ten thousand feet, on the Fourth of July weekend, 2008.

(Opposite bottom) After the fog lifted, this was the view from my campsite at Big Lagoon.

(Bottom) River otters check me out on my way from Big Lagoon up Maple Creek.

Cypress trees grow on the banks of the Edisto River near Givhans State Park at River Mile 60.

After a quarter mile of paddling, I reached the mouth of Pond Creek.

(Opposite top) A collaborative artistic effort set the tone for my journey up Pistol Creek in Blount County, Tennessee.

(Opposite bottom) A spillway with steps makes it easy to launch at Hematite Lake, a small reservoir in Land Between the Lakes.

(Top) My progress up Pistol Creek was halted by the pile of deadfall and trash behind a log that blocked the entire creek.

(Opposite top) These crossed trees mark the entrance to the maze at the back of Energy Lake on the Cumberland River side of Land Between the Lakes.

(Opposite bottom) My niece Libby accompanied me on a paddle up Hematite Lake.

(Top) The moon made an appearance as biologist Drew Crain and I exited the darkness of Citico Creek below Chilhowee Dam on the Little Tennessee River.

(Opposite) Sounds seem amplified and mysterious on a dark creek in the dead of night. Photograph by Drew Crain.

(Top) On an early spring paddle up the Powell River near New Tazewell, Tennessee, Norm and I decided to explore this quiet creek.

American toads mated on the bank of the Powell River in a configuration that Drew Crain said increased the chances of "success."

Norm wasn't keen on entering the cold waters of Calderwood, so I had to reverse roles and retrieve the ball I'd thrown.

Norm surveys the Little Tennessee River from the crest of Highway 129, known as the Tail of the Dragon for its numerous hairpin curves.

BRACKISH WATERS

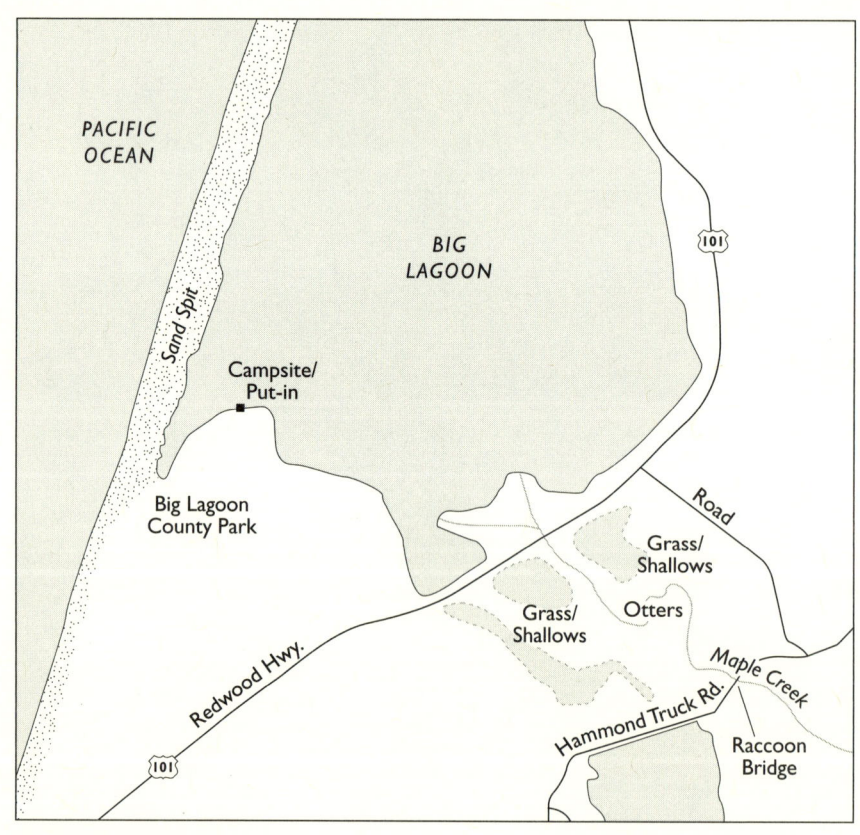

Big Lagoon/Maple Creek

CHAPTER 9

BIG LAGOON TO MAPLE CREEK: FROM ONE WORLD TO ANOTHER

When I was ten, in 1968, my father drove me and my mother to California to visit her sister, Emmy Lou, and Emmy's husband, Mike, who lived in Beverly Hills. We left western Kentucky in a 1965 aqua-green Oldsmobile Delta 88, a long, sleek car whose engine ran for 280,000 miles until I borrowed it in the 1980s and managed to set the engine aflame when a fuel line sprang a leak. This was the giant car in which my father taught me to drive, and after it had been in the family for about twenty years, I finally succeeded in destroying it. For the trip to California, my father had a AAA "triptik," which was a flipover map that had the legs of the trip on each page, with the preferred route highlighted in yellow. Part of my job was to consult the map and announce the upcoming towns and landmarks. Near Denver, at night, we somehow got lost. I remember dozing in the backseat as my father steered the Olds up mountain switchbacks long past midnight. My parents were quiet, tired, and frustrated. Once we got to California, my father complained about the aggressiveness of drivers and the lack of directional signage. After visiting my aunt and uncle, we drove up north to Monterey, where the weather was cold and damp, and no one swam in the ocean that crashed onto the beach. I didn't think I'd ever go back to Northern California; it seemed aloof, cool, and impenetrable, like the fog that hung in the air throughout the day.

Forty years later, I was driving down Highway 101 without much of a timetable, not really worried if I got lost or not. My only task: find a suitable campsite next to the ocean, the kind that Buck, the campsite host I'd met at Montana's Salmon Lake State Park, said was worth thirty-five dollars a night even without flush toilets. I wanted to sharpen that memory I had of the wild northern coast. It appealed to me now more than it did during the trip with my parents. I crossed the mouth of the Klamath River near where it met the Pacific and passed up Prairie Creek Redwoods State Park, where the Roosevelt Elk grazed near the road like cattle. I passed by Humboldt Lagoons State Park and Stone Lagoon until something told me I'd better check out Big Lagoon or I might be camping in an atmosphere that had been filling up with the smoke of more than eight hundred wildfires in central California. It was mid-morning when I turned onto the narrow road that led me to a county park of ten campsites, shaded with one hundred–foot tall cedars, dwarves compared to the redwoods I'd camped under the night before at Jedidiah Smith Redwoods State Park. Here at Big Lagoon, the fog lay heavy, and the wind blew chilly from the ocean, blocked off from the lagoon by a three-mile-long sand spit. I put my fifteen dollars in the box and parked at a site just across from a couple who were leaving. Only a couple of other sites were occupied, the people in tents or small trailers. On my site grew cedars and juniper with twisted trunks, testament to an intricate history of tortured growth. A few feet beyond the trees, on the beach, a white and yellow sailboat lay canted invitingly. The lagoon was nine miles in circumference, and by the time I got my tent up, the fog was on its way out, shreds of vapor lingering over the water. There had to be something wrong here; it was too good. I asked the girl who was leaving if she thought my site was better than the one they were leaving.

"Totally," she said. "You should camp there."

Her guy said they'd loved their stay, but the only "bummer" came from the fact that they'd set up their tent downwind of the campfire pit, so they got "smoked out." I set up my tent with the door facing the lagoon and meandered off to purchase some firewood to ward off the damp and cold that would descend that night. The guy cleaning the bathroom directed me to a small humpbacked trailer off by itself, pine needles, cones, and limbs scattered across its roof.

"Just knock on his door if he's not outside," said the bathroom guy.

A guy in his thirties answered the door, bleary eyed, hair sprung up, shirtless.

"Sorry to wake you," I said. It was almost noon.

"No sweat," he said, "no sweat. You want some firewood?" His voice croaked a little. He told me I could have ten split logs for five dollars, but I had to load them myself into the wheelbarrow beside the pile. When I returned the wheelbarrow, there was no sign of activity at the camp host's trailer. Asleep again. He had no golf cart, no stuffed bird that sang when you pushed a button, no canopy with bright lights, no walkie-talkie, and he certainly hadn't retired after a long career selling life insurance. Dude, he was working at the county park, living in the moment. A surfboard had to be around somewhere.

A ten-minute walk took me along the lagoon to the spit, a hump of black, gray, and tan pebbly sand and rock that formed a barrier between the Pacific, far from peaceful, and the lagoon, an expanse of lake-sized water that flashed electric blue when the sun bore through the fog and lit it up. In the open sea, waves crashed white, blue, brown, and amber, spreading out foamy tendrils that chased me as I walked along the beach. I scanned the horizon for whales. A sea lion surfaced once, then again. A couple walked past, bundled in hooded fleece from the waist up, shorts below, wielding what looked like golf clubs. They were hunting agates, small multicolored gemstones that wash up on the beach. Off in the distance walked one girl alone; farther away, hazy in the spray of the surf near the distant point, a man wearing an orange cap explored. I could have stayed there all day, inspecting each wave as it rolled in, kneeling among the ferns at the top of the mound, pale green fingers reaching up among tiny serrated leaves that grew close to the sand. A pale purple flower with tiny dark dots nestled among soft oval leaves spread itself out across the sand pebbles. Gulls, osprey, and pelicans circled the gray sky. Some perched on skeletal driftwood that lay about like dinosaur bones. Only my desire to paddle up the lagoon got me to tear myself away from the beach and the mesmerizing ocean.

On the walk back, I met Marna, who had a big white pickup and a trailer with racks full of kayaks that she rented. As her two dogs watched from under the truck, she instructed the lone girl I'd seen earlier on the basics of kayaking. The girl was seated in a boat on the beach fronting the lagoon. Marna wore a black knit cap, sunglasses, shorts layered over a wet suit, and a heavy plastic black knee brace, tiger striped, evidence that she'd done some "gnarly" paddling somewhere, sometime. She said she didn't mind if I listened in to the lesson.

"Rotate your hips to save your shoulders," she told the girl, who was a good student, quiet and attentive. In a half hour Marna had her at the water's edge, poised to embark on her first kayak voyage, alone, in the lagoon.

"Let me take my socks off," said Marna. Barefoot, out of her blue crocks, she held the sit-on-top kayak steady as the girl lowered herself into the seat. She winced.

"Now I told you your bottom would get wet, didn't I?" said Marna.

She smiled and nodded. Marna told her she could go out to the point of the jetty and back or she could paddle up the lagoon to the creek. The girl headed to the tip of the jetty, to the opening where the sea met the lagoon. Marna warned her to stay away from the opening, where the seas were rough, the undercurrents strong enough to pull her into the open water.

In the spring and fall, Marna told me, she paddled the Smith, where I'd come from, and the Mad and the Trinity. She told me about friends who had been injured in bad spills, daredevils who got paid to travel the world and kayak over waterfalls for filmmakers. One guy had compound fractures of each shinbone after a bad landing off a waterfall. I briefly summarized my Rogue River experience, blaming my blameless boat for the mishap and the unplanned swim.

She didn't laugh. "In white water, I want to be able to turn on a dime," she said. There was a place on the Smith River, she said, where if you fell out, you had to swim for a mile. When serious kayakers talk about "swimming," they mean trying to breathe air and not water and to avoid getting crushed against a rock or a fallen tree.

I could paddle a ways up Maple Creek, she said, but at a certain point it became the property of the Big Lagoon Rancheria Tribe, and they didn't like people getting out on their land and peeing, for example.

From my campsite to the lagoon, it was about ten steps for me to unshoulder the boat, drop it in the water, sit down and paddle. It was noon, the air still cool and damp, the sun making an occasional appearance through the fog. This would be a transitional zone different from any other, where moving water met the pooling of the ocean, a transition from salt to brackish to fresh water. The wind was at my back as I headed for the Highway 101 Bridge, the mouth of Maple Creek hidden by a field of green grasses with secret passageways such as I'd seen at the St. Joe River in Idaho. Once I was within the grasses, which were five or six feet high, the water smoothed out, and the wind stopped. My boat slid over underwater grass that grew in the narrow channel I'd chosen. Past the bridge the passageway widened to thirty feet. I could not see the bottom. Something caught my attention on the left bank. Four otters, huddled together, their gray fur slicked down, eyes on me, seemed passive in their interest. They sat there as I backpaddled toward them. Only when I was within fifteen feet did they scoot into the water and disappear, one by one. An outrider swam around my boat and poked his head up every few seconds in a different place I could not predict.

Farther up the creek I wound through the sea of grasses. There was some current, but the volume of water was so low, the gradient so gentle, that I could have paddled against it all day. After a mile or so, trees grew on the banks, and I might have been on a creek in Tennessee, the same kind of thickets on each side, water trickling over gravel bars. Under another bridge I coasted. I had to stop, no question about it, and I was sorry to trespass, but it couldn't be helped. As I got out on a rock outcropping under the bridge, the biggest raccoon I have ever seen moved from the weeds up the hill to the empty highway. Perhaps he belonged to the tribe: I asked his pardon as he ambled away, in no hurry. Above was a narrow paved road that looked abandoned, its origins obscure, destination nonexistent. A road to nowhere. As I paddled farther, the creek banks squeezed me in, then widened into pools like I used to fish with my friends on Clarks River, a tributary of the Tennessee (see the epilogue). Marna's warning about the reservation kept me quiet, and the creek had an intimacy about it that made me wonder if I'd come around a bend and surprise somebody doing something I'd never seen before. It was a little creepy, that creek, but I liked it, and I followed it upstream until fallen trees blocked any passage short of getting out and dragging the boat across yards of gravel bar and brambles. It was amazing that such a closed-in, intimate place could be so closely connected to the vast violence of the ocean, just a couple of miles at my back. The return trip was just as fascinating as I traveled from a place where trees blocked my view of anything but the river to the open-aired grassland to the lagoon, which rippled now with a strong wind that interrupted my contemplation and forced me to exert some energy to get back to camp.

The light turned golden and sharp as the sun sank to the brink of the sand spit that evening. A guy in a jeep pulled into a campsite down the way, set up his tent with precision and efficiency, changed into long pants, tucked in his shirt, put on a watch cap and a short green jacket, and strode with his camera toward the sand spit. This was the only lone camper I'd seen the whole trip, my alter ego, though he seemed a much neater, more organized version of me. Across the way camped an Australian woman and her son. For a while she was joined by a goateed man who drove up in a BMW, but after walking around with her and her kid, he left. "It's nice to be so close to town," she said to me after he left, "so that friends can visit." She sounded a bit wistful. I let the sound of the surf lull me to sleep, happy to have her nearby, though content to leave her alone, her wistfulness hanging in the air like the shreds of magical fog.

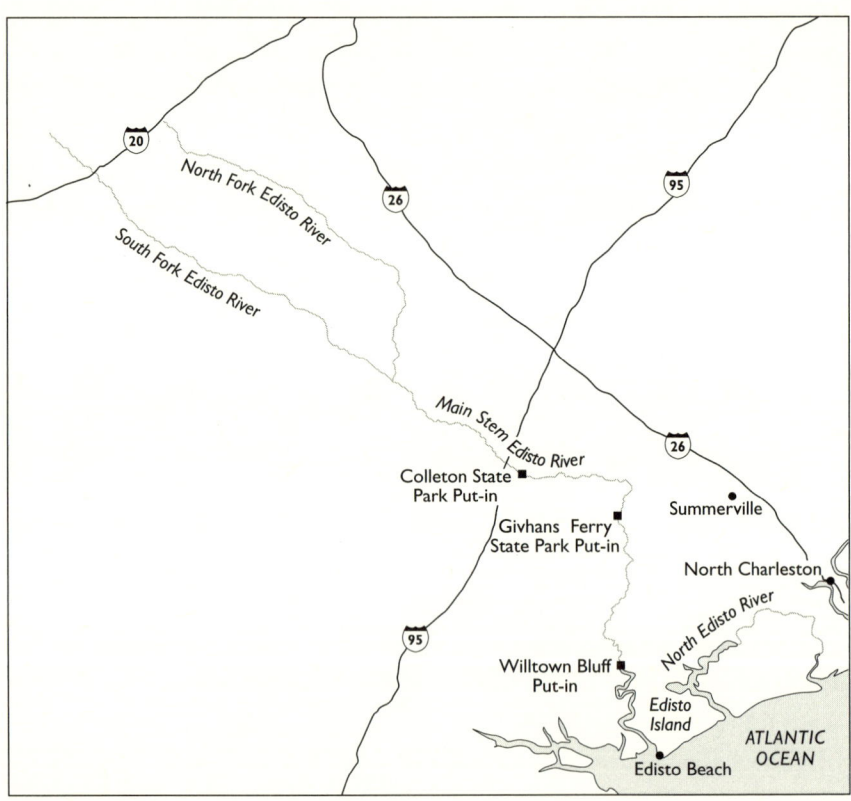

The Edisto River

CHAPTER 10

FEAR, DELUSION, AND PEACE ON THE EDISTO

I was driving toward Edisto Beach on a flat, patchy two-lane that passed through hamlets dark in the shade of live oaks, Spanish moss drooping in lacy clumps like spirits dropped from the sky. Modest frame houses, some on raised foundations, lined the road, the lawns mown and uncluttered, the houses in good repair for the most part. Up ahead a clump of buzzards met on business between a church and its cemetery. They feasted, their wings rustling, heads bobbing up and down over something big enough to serve an extended family of seven or eight. A less fortunate deer, which couldn't have been dead long, lay there like a banquet table among the birds, the object of their hunger. Something was going on inside the church as well. Traffic was heavy, tourism going full force on this road to the ocean that included, besides the dwellings, a serpentarium, a museum, a bookstore, and other thriving businesses. In the next couple of days of driving up and down this road, I saw fewer buzzards, but the deer stayed, of course, his remains consumed by lesser custodians of the wild. On the radio in my car, I heard this: the Gullah people, descendants of African slaves who lived in lowcountry South Carolina, passed babies over open graves to ensure that the spirit of the departed does not haunt the child. Margaret Washington, a Cornell University historian, was more specific about the context of this tradition in her article "Gullah Attitudes toward Life and Death." She said that when a mother died, the child was passed over the open grave to prevent the mother coming back for the child. The Gullah have preserved much of their African spirituality since they were taken by slave traders from Sierra Leone, Liberia, Senegal, and Guinea and converted to Christianity. To the Gullah, the spirit world is "unseen but not remote," according to Washington. "Inhabitants of the spiritual world were the guardians of life, appealed to in periods of crisis, such as illness, environmental disasters and the malice of others." Gullah do not clean and mow and decorate their graves, said the cultural historian through my radio. One should not disturb a grave, they believe. At the same time, Washington has noted, "Death was a journey into the spirit world, not a break from life of earthly beings." Working in the brutal conditions of the rice plantations, "slaves lived in the presence of death constantly, and seemed to feel that the phenomenon was as much a part of living as their continuous labor. . . .While the Gullahs' perception of life after death was essentially of Christian origin, many practices associated with the dying and the dead were derived from African antecedents."

In this context the shocking but fascinating image of the buzzards and the deer began to make sense. This, so far, was an entryway into a different kind of tour, a journey to a place where the natural forces of death and decay were out in the open, where a radio program proudly told me about traditions that some might label superstitious or profane. On the eight-hour drive from East Tennessee, I had entered a world that made my heart beat faster, even in the small vacation community of Edisto beach, where things were a bit more sanitized. My friend Drew Crain, having his annual family reunion at Edisto Beach, graciously put me up for a night before I began my excursions upriver from the brackish to fresh water Edisto River.

My wife, Julie, had left me a note on the kitchen counter that said, "Be Careful," under a sketch of a toothy alligator. I drew a stick figure (me) perched on the bow of a canoe, a paddle extended toward the alligator's jaws, as if to prop them open, something like Tarzan or Jungle Jim might do: "Don't worry. It's under control," I think I wrote. I laughed about it then, when I felt confident that if I followed the standard code of behavior around predators—take a photo with a long lens and move on—I'd leave coastal South Carolina with all my appendages and dignity intact. That was before I passed through the threshold at the place of the dead deer and feasting buzzards. A day later, as I launched at a dirt boat ramp at Willtown Bluff, twenty-five miles inland from the beach where I'd seen dolphins breaching the water just offshore, where Drew took me on a night beach hike in search of loggerhead turtles burying their eggs, the idea of alligators was on the verge of becoming concrete, and it wasn't so funny. Drew, a biologist who knew the area well, hadn't heard of Willtown Bluff, and apparently neither had anyone else, because there was not another person in sight when I lay my yellow kayak down in the still water next to the green grasses that poked up like needles through the surface. "You *will* see gators," said Drew the night before, as we sat on the screen porch of the house he was renting, the breeze blowing in from the ocean. His brother-in-law, an FBI agent who grew up in coastal South Carolina, described water skiing as a boy in areas thick with alligators. I would meet other people on the trip who dismissed the danger of gators, including a Floridian family who said they were "no big deal," this while the guy was snorkeling with his kid neck deep in the blackwater swamp, harvesting river rock for Mom's kitchen.

Now, at Willtown Bluff, I paddled with stealthy strokes toward the bend in the river a half mile away, chest-high grasses on each side, a pelican perched still as an ornament on top of a wooden post. Up ahead fifty yards floated a couple of dark logs, which seemed kind of strange since I didn't see any trees on the banks. The logs seemed to be moving, though there was no current here, the estuary still like a lake. Odd. A trilling kind of dread arose from my bowels, from my vital organs, separated from the water by what suddenly seemed like a wispy plastic shield, a fragile boat of gossamer, my ass an inviting target for a submarine attack from these lurking monsters. I considered turning back. Now there were three, then four. I had imagined the South Carolina alligators to be lazing away in the sun on the bank, blinking sleepily as I took their pictures. In guidebooks I had read the phrase "sunning themselves on the banks" more than once. Never did I imagine them in the water with me. Moving. This changed everything. When I looked back and saw a man standing at the boat ramp straddling a bicycle, I paddled toward him, wondering if I my launching here had violated some law, some custom, whether I had trespassed into a forbidden zone. Casually I began the conversation with a fishing question. He'd caught catfish here, he said, big ones. There were also sturgeons in here, he said, but they scared him, and he cut his line when he caught them, even though biologists told him they would pay him for specimens.

"There are people who eat sturgeon," I pointed out.

He knew that. "They skin them like a catfish."

Still he wanted no part of this prehistoric looking fish with his rough skin and its teeth, very gator like.

"Are there gators in here?" I asked. "I thought I saw some."

"Probably are," he said, as if I'd asked about mosquitoes or flies. He had to go to work at one of the houses up on the bluff, so he turned his bike and left me there to contemplate my next move, whether I would load the boat onto my car and count myself lucky to have made a prudent decision or shrug my shoulders and adopt the attitude of people who lived among the giant lizards: no big deal. I paddled outward, hoping to find the place on this day where the Edisto narrowed to something that looked and moved like a river. I was in the ACE Basin, part of a system of wetlands and tidal creeks formed by the Ashepoo, Combahee, and Edisto Rivers. As the *Edisto River Companion* states, this is one of the "largest undeveloped estuaries on the east coast," 350,000 acres where "differentiating between land and water becomes difficult."

The gators were moving across the river like a squadron. I hugged the right bank and paddled past them at what seemed a healthy distance of twenty or thirty yards, though who knew how fast they could swim with those powerful tails, those thick legs with the long claws—surely razor sharp, one swipe enough to cleave my boat in two. (Alligators can run on land up to nine miles an hour, though no one seems to know how fast they can swim. Their cousins, crocodiles, can swim ten miles an hour, about twice as fast as I could paddle.) Hoping to sneak past them and around the bend, I began to wonder if they would form some sort of blockade between me and the ramp, for there was no place anywhere to pull up on the bank. There really wasn't a bank, only the ominous, inscrutable fields.

Only an hour before, I was congratulating myself for making the effort to launch here. I'd found the ramp on my own, driven down a road that turned to sand, the live oaks reaching out over it, forming a tunnel through which I had passed, the morning sun illuminating the entrance behind me. Near the ramp was a majestic nineteenth-century plantation house with several brick columns that framed a wraparound porch. Willtown, a sign told me, was established in 1685. There had been a Presbyterian church here.

Now, after passing the gator gauntlet, I was still considering aborting my quest. Who knew how many I'd see as the river narrowed and we were forced to share tighter spaces? Later, I would read things about alligators that I'm glad I didn't know at the time. Their mode of attack would be to lie very still and wait until a small animal (often a pet such as a dog) or sometimes a child would wander up to get a drink, then the alligator would spring forward, clamp down the jaws, hold the victim underwater until it drowned, then swallow it whole. Don't swim at night, said the travel books. Don't feed the gators. Keep your distance, it said.

Like Marlow heading up the river to fetch the crazy ivory trader Kurtz in Conrad's *Heart of Darkness,* I kept going up the Edisto that week, leapfrogging from car to boat, seeking out the primitive dread that arose within me in the proximity of large reptiles, the fear itself something that linked me to them, something I cherished and cultivated as the river grew more and more narrow and I hoped I would finally come face to face with one of the monsters and get the shot I was hoping for without being consumed, camera and all. That first day, at Willtown Bluff, I experienced the novelty of my virgin voyage with gators, and like Marlow on his first visit to the Congo, seeing up close the horrors of colonialism, I was shocked by the inscrutability of the place, the calm water, the blank, inaccessible shoreline of grasses, me and the death barges the only things moving. I paddled another two miles, stopping when the wind rose in my face and ruffled the water's surface. Breezes troubled the fields of grass all around. I headed back to the ramp. Not another gator did I see that day, and, in the assurance that grows from surviving a trip into unknown waters, I drove upriver to Givhans State Park, at river mile 60, for my next sampling of the Edisto, confident I would see the stereotypical alligator blinking in the sun and maybe some kind of bird companion perched on his back. By now, after so many reality shows about people catching giant catfish and turtles and alligators by hand, the actual prospect of being there, in a boat, near an alligator, may have lost its punch.

At the state park I felt, to an extent, that I'd passed back through an invisible threshold to land a bit more familiar to me: the campsites with the little gravel strip to park on, the trees familiar types without the Gothic hanging moss, the shower house, and the administrative headquarters with the friendly ranger and the brochures. You had to descend a path of fifty yards or so to get down to the river, where people fished and swam and rode rubber rafts in the slow current. Here's where I got the first sense of the term *blackwater,* in actuality a misnomer that led me to think that the water was polluted and stinky and stagnant. It also triggered an unfortunate memory of a Doobie Brothers song that was overplayed on my high school jukebox during lunch period. The Edisto here, near huge swamps with fresh flowing springs, was the color of strong iced tea, and though you couldn't see to the bottom, you could see a few yards. The water, which does not stink, is dark because it flows from cypress swamps that yield tannic acids. According to the *Edisto River Companion,*

"These acids, mixed with decaying leaves and sand, give the river a dark look," and "blackwater rivers are some of the cleanest natural waters in the world."

I planned to paddle upstream as far as I felt like it and then float back down. Again I was pretty much alone in my quest, and the current was so slack that I almost felt I was paddling on a lake. I passed a couple of houses sitting back at the crests of hills, including one where a dock was being built. After four miles of paddling past the beautiful cypress trees that grew from the water near the banks, their intricate trunks (knees) mirrored on the water's surface, I reached a group of houses on the bank, what looked to be a small community; the river narrowed and seemed more resistant, so I turned and let it carry me back. At the mouth of Four Hole Swamp, I stopped and chatted with a man fishing from a pontoon. He wore a straw cowboy hat and a camouflage t-shirt. He had fished the river for forty years, he told me, and, like most people who were attached to a beautiful place, he mourned the loss of its purity over the years. It was still pretty clean, he said. If you want to see a gator, he said, "there's one that lives behind that tree over there and comes out in the evenings." He talked about this lizard like he knew him; I was expecting a name. Not far from there, a father and son snorkeled. I headed a half mile up Four Hole Swamp, under a highway bridge, and then headed back to camp. I had no idea that I was at the threshold of a sixty-two-mile-long swamp fed by a freshwater stream, the largest tributary of the Edisto. I was there long enough to realize that swamps, like creeks in other parts of the South, were secret places full of surprises, and I was motivated to find another one once I moved north to my final segment of the river at Colleton State Park in the heart of the fifty-six-mile-long Edisto Canoe and Kayak Trail. I still hoped for a close-up photo of an alligator.

At Colleton State Park, I belonged, in a sense, and unlike Marlow, whose upriver trip took him farther and farther into the jungle and the insanity that Kurtz represented, Colleton, unlike Willtown Bluff, was a place of river lovers and paddlers. I should have fit right in. As often occurs to me in places where I'm supposed to belong and fit in, here at Colleton I felt something close to despair, at least when I first got there in early afternoon. Part of it was the melancholy that arose from the people there. First I stopped at a canoe outfitter across the river from the campground. I had to rouse a young man, Sean, from somewhere inside the store. No one else was around, nobody wanted to paddle the river, and perhaps most melancholy (and comic) of all at such a deserted business was a canoe that must have been fifty feet long, set out in the parking lot. Sean said it was a working canoe that they took out, that it would hold thirty-odd people. He offered to shuttle me for a small fee if I wanted to go downstream for a day or two, and when I told him I wanted to find a place to paddle upstream to the tree houses that were rented to paddlers, then float back to my put-in, he seemed baffled, then dubious that it could be done here, where the main stem of the Edisto had narrowed and quickened considerably. These kinds of statements tend to motivate me. I took my leave of Sean, and he went inside to continue his nap.

Across the way, the campground was far from full, but pretty well populated with campers, swimmers, fishers, and paddlers. I picked the most secluded spot, deep in the shade of small trees, a low place where you could take a path to the river and a wooden walkway built out from the bank. Standing there that afternoon, I heard before I saw a woman who must have been a boy's grandmother lecturing him all the way downstream about the way things used to be, how lazy he was, and how much backtalk he got away with. I then spotted her: wearing a white visor, she was seated in a black inner tube, more in the water than out of it, and she pushed along a big blue and white inner tube with its own cooler. That night, after I got my fire started and had settled in for an evening of flame contemplation, a guy came over from the campsite next to me and said this: "Permission to approach?" He wore a nylon rain hat, hiking boots, and shorts; everything looked like he'd bought it the day before, all squeaky clean and unwrinkled. His group already was raising hell, and I had been able to hear the conversations. I predicted a fight, at least some hurt feelings before the night was over and they all passed out. But this guy, who was sober, had appointed himself as some kind of ambassador to come over to me and explain that they would be partying that night and if they got too loud to just let them know. Odd, I thought, that he didn't at least extend a perfunctory invitation to join the group for a drink. I brooded for a while, and then called my friend Tim Parrish from my days at the University of Alabama twenty years earlier. He didn't answer, but I left him a long message in a staccato, overly enunciated voice of a type he often mimicked, the overly serious graduate student, and I said my name was Jimmy Kelso, and I needed some help in researching my dissertation. I laughed to myself like a madman after the call and then slept well in spite of revelers on each side of me. (Tim would wait a year to call me back and use the Jimmy Kelso moniker.)

I was rained in for much of the next day, but with the help of the ranger I discovered a put-in at Green Pond Church Landing, off Highway 61, at the end of a five-mile-long county road. He was a bit worried that the landing, which was actually on a creek, wouldn't have enough water in it for me to paddle the quarter mile to the main stem of the river. At the landing, a dark pool festooned with bright-yellow water lilies, roosters crowed, and thunder rumbled from upriver. I barely had room to turn my boat around in the tight quarters, though at some of the wider openings I would be able to cast my fly rod and tempt the famous Edisto red breast with small floating poppers I jigged along the surface. The swamp rose all around me as I skirted under and around deadfall, the place dark and quiet like some kind of outdoor church.

I kept my paddle strokes light, certain that on this, my last day with the Edisto, surely I'd come upon a gator at closer quarters than the Willtown Bluff flotilla. I fished as if oblivious, a decoy for the gators, but I kept my camera and my long lens ready. A movement of something behind a tree not much more than ten feet away made me flinch and squint my eyes. It was a mama raccoon with three babies foraging among the exposed roots and brush, in no hurry to back away as I approached and passed through the narrow opening. I came to a pool open to the sky and fished for a while. There was no action other than the silent patter of raindrops on the surface, the thunder rumbling in the distance. How far I'd come from the trepidation of that first day on the Edisto, when I almost aborted my voyage in fear of the alligators. Here, on this swampy creek, where the natural world loomed on all sides, I felt closer to this place, more at peace than I had been anywhere for a long while. The Gullah, like many people, believe that the dead cross a river on their way to the afterlife. Here, feeling this calm as I cast my fly rod, having emerged into the pool of light from the darkness behind, I believed it literally, though I didn't think I was dying or near death, perhaps more reconciled with mortality than usual.

After a few more paddle strokes, I was on the main stem, working against the current on a river much more narrow than downstream at Givhans Ferry, only about thirty feet wide at one bend, where it was further constricted to ten feet by fallen trees. I paddled hard less than a mile before I found the tree houses I'd read about, blending in with the woods on a big island. They rented for $125 a night, and I treated myself to a free tour. You climbed three flights of sturdy wooden stairs to the shelter, which had a working kitchen, a cushiony couch, a table covered with a white cloth, a loft with bunks, and a deck, which included a grill. You were in the forest but also in some ways above it, at a vantage point that afforded a commanding view of the river and the sky. As I prepared to head back up the creek to the landing, I heard something substantial moving through the woods toward me. It turned out to be a young woman named Weatherby (her first name), who worked for the outfitter REI. She was guiding a big group down the river and had arrived in advance to prepare the shelters. Having grown up in Edisto Beach, she said, she'd seen plenty of alligators. They were around, she added, but you usually hear them before you see them—they make a grumbling noise, and then you hear a splash that is "unmistakable."

I paddled back up the creek to the landing with a renewed sense of awareness about gators and any other wildlife that lurked in the ghostly landscape. Though I would never get the gator shot I wanted on this trip, just looking for them and being aware of them in the quietness of the swamps seemed to intensify my experience. That the Edisto was the longest free-flowing blackwater river in North America made me glad that I'd sampled it and got a sense of how it transformed from the mouth near the ocean to the swamps upriver. Though I didn't find an exact point for its transition from brackish estuary to free flowing freshwater, it seemed that the Edisto, in some form, offered a liminal experience every day I paddled it.

DAMAGED WATERS

CHAPTER 11

SEEKING DAMAGED WATERS

I don't confine myself to the pure and unadulterated in rivers—"scenic," "unspoiled," or any of those other adjectives that we use to separate the idyllic from the homely ones, those rivers we've judged less deserving of being set aside and protected. At times, particularly on trips out west, it was a relief and great comfort to find a place as secluded and cold and clear as Platoro Lake/ the Conejos River in the San Juan Mountains, where I spent a quiet Fourth of July in 2008 (see chapter 8). But often, less than pristine rivers are just as interesting, if not more so, than the beauty contest winners. Foremost, someone's got to document the trash on a river, to make people aware of what's out there that is visible from the intimate perspective of a canoe or kayak on the theory that a sustained cleanup effort will follow the awareness. On our 2003 trip down the length of the Cumberland, Randy Russell and I saw sections of unexpected beauty on the first one hundred free-flowing miles in eastern Kentucky. We also saw pipes that pumped raw waste into the river. Alongside the deeply wooded shorelines and the sheer rocky bluffs, amid the roaring rapids, flashing white, we cataloged discarded or misplaced objects unique to our species: appliances such as washing machines (no dryers!) and televisions; dozens of vehicles half-submerged, hood first, as if pausing for a drink from the river; vestiges of construction, plastic cones, broken concrete, lumber, and some kind of fabric waving like curtains from trees for miles; the usual detritus of recreation, plastic and glass bottles, tires, lawn furniture, all of it evidence of our energy in acquiring and consuming as well as of our laziness in disposal— crazy stuff such as the toilet seat hanging suspended in a tree on the lower river in western Kentucky, a memorable vision for the last day on the river. I documented this in my second book, *Coldhearted River: A Canoe Odyssey down the Cumberland,* feeling it was my duty to give a complete picture of the river, not just the pretty parts. In 1996 Vic Scoggin, who grew up near Nashville on the river, had swum the length of the Cumberland. What he saw and felt and smelled and tasted was no doubt worse: he mentioned raw sewage, dead cows, and chemical and industrial waste at a level that convinced Randy and me that there had been some cleanup since Vic's trip.

Apart from the need to document damage that can be perceived by the senses—what any ordinary paddler can do, if willing to embark upon a damaged river—I also must admit a more detached interest in the phenomenon of trash: how it gets into a river or onto its banks and what it says about us. What is the process by which cars arrive at river banks? If they still run, why abandon them there? If they don't run, wouldn't it be a lot of trouble to deposit them in these remote locations, far from roads? What is the thought process that takes place when someone who fishes, swims, or boats a river finishes an Egg McMuffin and drops the greasy wrapper into the water? Or, more commonly, finishes a cigarette and deposits it like sizzling bait onto the surface of a river that was a childhood companion? Don't tell me that those folks just aren't thinking. You have to be conscious to eat, drink, and smoke, and you have to cognizant of the deliberate decision to add visible advertisements of your consumption to the flora and fauna of a lake or river. Some items defy explanations and imagined scenarios, such as the bowling ball in the Cumberland near Burnside, Kentucky. Did you know that bowling balls can float?

Paddling damaged rivers, these smaller questions lead to larger ones. How does a coal-fired steam plant affect the experience of paddling a river or lake? Jasper and I, on our Tennessee River trip, camped next to a bay warm as bathwater, smelling like wet, dirty rags, as a result of the plant's discharge. By now we also know that retention ponds holding coal ash can fail and result in epic messes on rivers, case in point the Emory in Roane County, which underwent a massive cleanup that has taken years. I paddled a portion of it when it "reopened" in 2008 and was amazed at the transformation compared to the previous mess I'd seen in videos and photographs. Impressed by what the massive cleanup said about our capacity for correcting mistakes, it also made me wonder why the same kind of energy and ingenuity can't be channeled into prevention and vigilance. I also wondered what vestiges of the spill might not be evident to the senses of an ordinary paddler.

Going upstream toward transitional zones slowed me down and gave me the leisure to contemplate in greater detail the impact we have on waters. It made me realize that just about all waterways are damaged in some way by our use, and it caused me to consider how the effects of our industry and recreation on lakes might differ from our impact upstream on the moving waters that feed them. My assumption was that the lakes would improve the closer I got to the transitional zone. That was not always the case.

CHAPTER 12

UP PISTOL CREEK

Because they rarely show their power unless engorged by heavy rains, urban creeks often hide from citizens' sightlines. Sometimes they are all but obliterated by channelization and concrete, the trees and brush along their banks stripped away, their lifeblood rendered stagnant and stinking with the runoff and discharges of human activity. Which is a shame. Because sometimes a creek pumps a town's lifeblood. Pistol Creek flows through the small city where I live, Maryville, and the town next to it, Alcoa. The paved Greenway follows it through three parks, where people transverse it multiple times on wooden bridges: walking, running, and biking. Kids and dogs wade in it. Geese, frogs, toads, fish and ducks live in it. I haven't seen any boats on it, but sometimes, after heavy rains, sections of it look like class IV rapids, with foaming deadfall dams and narrow bends where water crashes against the banks and under bridges. I've seen it overflow many times, and I've seen it nearly dry. In a culture where a creek's main purpose seems to be in the naming of subdivisions or strip malls or liquor stores, Pistol Creek seems a mainstay, a high profile survivor, unlikely to lend its name to commerce. I'd seen quite a bit of the Pistol from above, jogging on the Greenway, so I decided to explore the town from below, in my kayak, to see what the ducks and fish and dogs and kids had discovered.

    I've seen urban creeks much worse off than Pistol Creek, which seemed to me, at least in terms of what I could see from above, fairly clean, despite its course through two small cities. Other creeks, in nearby Knoxville, for example, aren't so fortunate. Sometimes, just to see my nature-writing students' reactions, I transport them to Knoxville's Third Creek, which flows into the Tennessee River not far downstream of the University of Tennessee's Neyland Stadium and the wastewater treatment plant. I don't tell them where we're going or what we'll see, just that it's urban hiking. This comes after a few hikes in the Great Smoky Mountains National Park, more of what they're used to when they think of nature. As soon as we get out of the van, the students see the white signs with the black lettering warning against contact with Third Creek's bacteria-infested waters. They usually make noises of disgust and might protest a bit at having to hike there. A fascinating variety of detritus decorates the banks and floats along the current as Third Creek runs through Tyson Park under an interstate and disgorges its contents into the Tennessee. I tell them to keep their eyes and ears open, to sniff the air. There used to be a factory on the far bank, a bellows with a tall brick smokestack and pipes that drained something or other into the river. There were smells that resembled the flatulence of industry. Now there's only a football-field-sized concrete slab. The irony of it is that on Third Creek we would often see more wildlife than in the Smoky Mountains or some of the state parks I take them to. Great blue herons, ducks, hawks, groundhogs, and a variety of songbirds are active along the corridor, seemingly at home among the highway construction barrels, the cans and bottles and various containers that descend from the neighborhoods and the busy highways.

Once I kayaked down the Tennessee River to Third Creek before a football game between Tennessee and Florida. Third Creek was my refuge from the Vol Navy, an endless procession of cabin cruisers commuting to the game. At half throttle they raised wakes big enough to break white and crash against the riprap-enforced banks with gale force. Up the creek, away from the hubbub, it was eerily peaceful, and I was able to paddle about a mile before I saw a large green head and torso floating in the water under a canopy of low hanging branches. I paddled forward, curious but dreading what I would have the misfortune to discover, smell, and report to the authorities. A drunken Tennessee fan who had stumbled and fallen into the creek? A Floridian who was pushed there? A homeless person who drowned himself? None of the above. Turns out someone lost their life-sized Shrek; the little horns gave him away, his smile frozen in idiocy. This was in October. In January, when I took my class there, Shrek was still decorating the creek, still smiling. What had been his path to this fate? Did some kid's big brother throw him out the window from the interstate in a pique of revenge for little brother's crossing the backseat boundary line? Had a group of city kids, deeming Shrek passé, buried him "at sea"?

During the Florida game, a lull in the big boat traffic allowed me to escape from my dubious refuge and make a break across the river to the less developed side, to Goose Creek. There a young man and his three children were fishing from a small concrete highway bridge in the deep shade. There was just enough room for me to cast small poppers with my fly rod and catch a blue gill in front of them. The man asked me if I knew the score of the game, but he didn't seem to care that I didn't know. Sometimes we would hear the crowd roar and the PA announcer's strident nasal narration, but it seemed a world away from us in the shady little creek. Goose Creek looked good here, at its mouth, but farther up I'd done some trash pickup during the annual Tennessee River Cleanup, and we'd found enough discarded tires to stock a retread store, plus a variety of other disturbing trash such as an automobile gas tank and a slew of unpackaged syringes. So, urban creeks are welcome refuges from crowded open waters and noisy, hot city streets and commerce, but at the same time, life above sends its worst down below. I was hoping Pistol Creek would resist this stereotype, that it would be as idyllic a voyage below as it was above, where it burbled and flashed white as an accompaniment to exercise, picnicking, and romantic strolls.

Early one July morning, I put in at the Rockford City Hall, a building the size of a one-room schoolhouse, just a few hundred yards upstream of a low head dam at the textile mill. Upstream of the dam, the Little River, which flows out of the Smokies, lay motionless like a narrow lake. Paddling the slack water was like reacquainting myself with an old friend, the sky casting vivid reflections of standing trees, fallen trees projecting like jagged arms from beneath the surface, their twins reaching out below, a parallel reality. The Little River was about fifty feet wide here, open to the sky. On river right a small gap appeared in the lush wall of intertwined sycamore, dogwood, and maple foliage. It looked like a dead end, the mouth of the Pistol, because it slanted so far back to the right away from the sightline of a paddler approaching from upstream, as I was. But once I entered the mouth, the creek opened up into what one might have called a fort, as a child, or as a working adult, one might think it the size of a classroom or a convention hall, big enough for a hundred canoeists to have an impromptu powwow, dark and subtle like a cave, the canopy of trees reaching over it and blocking out all but a few darts of sunlight. The temperature dropped, and the sounds of the traffic were muted, as if I'd entered an auditory buffer zone. Only the distant sound of a freight train whistle reminded me that I was near a couple of towns. There were no trails bordering this lower section of the creek, no easy access for the general public. I zigged and zagged slowly between fallen trees, taking my time, in no hurry at all. The liminal zone for this one would be the point where I could go no farther, where I surrendered to the creek. Having no definite destination, no timetable but my own tolerance for adventure and exertion freed me to pay closer attention to my surroundings. I hoped it would continue like this the whole way, an intricate course with easy obstacles, plenty of water to float in, and an absence of menace. I should have known better.

I picked a day after some moderate rain, thinking that I could make it upstream of the mouth for what looked like about three miles, to the duck pond at Springbrook Park in Alcoa, a good stopping point. This, it turned out, would be a wildly ambitious goal. I had no idea what obstacles awaited me. I could have scouted much of the creek from the ground, but I chose to let the Pistol reveal itself to me on equal terms, me in the kayak at water level.

The first bridge I passed under, Williams Mill Road, looked, from above, in a car, a bridge like many others, utilitarian and featureless, but underneath it was a sort of outdoor book, covered with graffiti messages. Some seemed innocent, even optimistic: a figure with bushy black hair and a red-checkered shirt with her hands in the air, a smile on her face, eyes thin lines that I took to be twinkles. Another smiling face, this one buck-toothed, curly strings representing hair. A big red peace sign. A happy face above a sad face, each scribbled by the same minimalist artist. Others, like the KKK sign and some threat to the "Po-po" (police), seemed quaintly menacing, the threats in their secret place unlikely to have much of an audience to frighten or offend. Others were so commonplace I wondered why someone would bother: why write "Kilroy was here" with the peeping face for the billionth time? Some seemed ill-considered or spontaneous because of the grammatical errors: "The Looses Where here." To wade through the snaky weeds down a steep bank and carefully craft a message with such egregious errors seemed absurd. Why wouldn't the artist write out a draft beforehand and have some more advanced artist proofread it before spraying on the real thing? For the first half mile, the painted bridge was about the only sign that humans had been in the vicinity, though I did come upon two basketballs spinning in eddies on either side of a deadfall dam, one a faded orange, the other black with blue lines. Again, one wonders what fateful occurrence brought about this unlikely scenario of detritus. What other strange phenomena awaited me in the depths of Pistol Creek?

I paddled maybe a half mile with no problems, ducking under a fallen tree and skirting a house-sized pile of deadfall that blocked all but a passage a yard wide. The creek offered no resistance in the form of current or shallow water. All was well. Then, after a quarter mile or so, the creek and its inhabitants began to intimate warnings. An enormous great blue heron rustled the limbs above me. He didn't squawk and fly upstream like they usually do when disturbed. He hovered above me in the small limbs that met in the middle. There was no nest. After I'd passed under and beyond him, the bottom came up, and the current quickened, what I took to be Pistol's liminal zone. I was glad to see the little creek come to life, but I had to work hard to keep the bow straight and my momentum going forward. Up ahead the bank opened up on both sides, some kind of clearing. On the bank opposite the clearing, two white fowl with sloppy red masks (Muscovy ducks) stared at me from a log. They did not quack. Above them, in a pen behind a small house, stood a dog the size of a pony, his coat buzzed for the summer. He did not bark. Around the bend was what looked like a gushing waterfall and more shoals flushing a stronger current.

When the current became so strong and the creek so shallow that I could not make progress, I drifted backwards down toward an eddy where I could get out and survey what was up ahead. Going backwards, my heart began to pound. Not that I would have minded turning over or having to step out of the kayak, but I had my bulky Nikon out of the dry bag to photograph the crazy-looking ducks, and I loved this camera like a brother; it was my constant companion on trips, and if I dropped it in the water, I could never afford a replacement. As I drifted backward and got lodged between rocks, the pony-sized dog barked once, the ducks flapped their wings, and I imagined a large group of humans gathering on the bank in time to watch me flailing in the water among the rocks and branches, my camera sinking to the bottom and the kayak continuing on back to the Little River without me. But the creek seemed to take mercy and allowed me to blunder through the shoals and into an eddy and small shelf of flat shoreline where I got out and sank ankle deep in cool black mud. After I labored up the bank, I surveyed about an acre of what I would call a wasteland. Power lines ran in two directions above this field, intersecting near the middle. Dead slashed vegetation lay across the entire area; stalks and pole-like limbs crunched and rolled as I walked across them. Nothing grew here but small green shoots that resembled poison ivy. As I tiptoed across this in my river sandals to investigate what had created the whitewater that turned me back, I couldn't help thinking that if I were a snake, I'd love to sun myself under one of these poles and reach up to bite some pilgrim's bare ankle. I looked down as

I walked, trying to think like a snake, and from the shade another dog stared without barking, this one a light brown sheepdog mix slightly smaller than the pony dog on the other side of the creek.

In my short voyage I'd passed under Williams Mill Road twice, but it didn't register with me until now that there might have been a mill at some point and that dams are commonplace structures near mills, where people used to harness the power of a river to turn a wheel and grind corn into meal, for example. Sure enough, a short walk across the slashed field revealed a stone wall with a little gap where Pistol Creek raged through with the force of a class IV rapid, the creek forming an oxbow here that even without the dam would have strengthened the current. The mill dam was hidden, and you'd have to be looking for it on land to find it, but it was an imposing structure, the gray concrete block wall a foot and half thick, fifteen feet tall, with vibrant green moss splotched on top.

I had a decision to make: float back down the peaceful half mile that I'd "discovered," fish a little, and be home before the sun rose above the tree line, or portage around the dam and forge onward. I'd gone to the trouble of getting up before dawn, I thought, and the dam ensured that the waters above it would be deeper slack water, so I began the arduous process of carrying the fifty-pound kayak across the slashed power line clearing for more adventure. Again, I thought of snakes or holes or irate landowners or nonbarking dogs that bite as I worked my way across the uneven ground. As I watched where I was putting each step, I noticed a strange thing: golf balls, dozens of them, half-buried like the eggs of some prehistoric beast. Across the creek was a parking lot for people who mostly used this farthest reach of the Greenway for a bike ride or a hike. Turns out it was also the launching pad for lunch-hour golfers.

It was about a hundred-yard carry to a steep muddy bank where I could slide the boat down. Across the creek, in the parking lot, someone sat in an idling car and stared through tinted windows as I poked the weeds with my paddle to warn the snakes that I was coming their way. I eased into my seat, feet heavy with mud, and continued my expedition.

The creek straightened and narrowed to ten or fifteen yards, and the massive trunks of sycamores, oaks, walnuts, and maples leaned out over it, all of them seemingly designed for rope swings. But there were no rope swings. Above was the concrete walkway and nearby was a spur of the trail that led up to Clayton Homes headquarters. I paddled hard against a quickening of the current upstream of the imposing piers of Pellissippi Parkway, a four-lane, and the Pistol flowed faster here over shallow water. I paddled under two other bridges, one a small road, the other a footbridge. At a blockage of deadfall that would have required another difficult portage, the current twirled a basketball among an eddying crowd of plastic and glass bottles. I turned back.

Pistol Creek, I found out later, was on a list of creeks and rivers that the Tennessee Department of Environment and Conservation (TDEC) had designated "Impaired" in 2008, a condition brought about by "loss of biological integrity due to siltation," the root source being storm water runoff. Seven and a half miles of the thirteen-mile creek were impaired, according to TDEC. The list of rivers and creeks is a long one, and it of course includes the infamous Third Creek in Knoxville, which TDEC classified as "Category 5 Impaired" with a "water contact advisory due to pathogens." Almost twenty miles of Third Creek are in this category.

Compared to the three-mile voyage upstream, the float back was almost effortless, giving me the leisure, in the gathering heat, to snag a couple of blue gill, the small tippy boat magnifying their size and fight. At the portage a couple of new dogs came out of the woods and alarmed the neighborhood with their barking, but on the entire float I saw no one else on the water or on the banks but the person in the tinted-window car. Pistol Creek is by no means pristine, particularly near roadways, but exploring it was worth the effort, a change in perspective from the view on paved walkways: quieter, cooler, and full of surprises. Liminality here took the form of an unforeseen obstacle, an obvious landmark I should have known about that kicked me backwards and forced me to realign my preconceptions of the creek and the trajectory of my voyage.

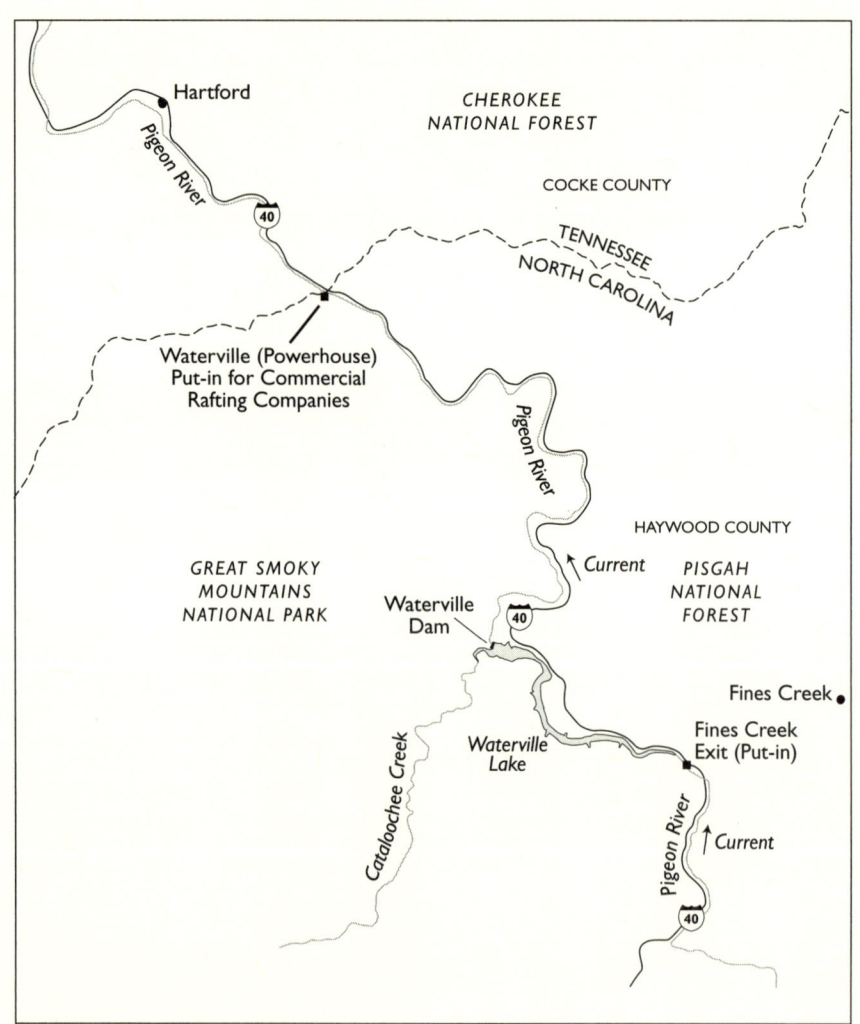

Waterville Lake/Pigeon River

CHAPTER 13

FINDING AND SMELLING THE PIGEON

You pull off the interstate onto a hidden road that leads to a clearing where they store salt to melt ice. It's June 2009. There is not one, but two barred gates across the path through the woods down to the lake, Waterville or Walters, you've seen it called both. The power company's signs tell you that walking the trail would constitute trespassing. You stroll past the gate a few yards, just to see how it feels. It's not so bad to break a foolish law, but to get caught would be humiliating, and leaving your car here for a day in order to kayak this lake would expose you to the minions of the power company, in cahoots with the local law, you imagine.

For years you've admired the river that feeds this lake. Every time you drive east of Knoxville on I-40, you creep in the slow lane and crane your neck to see the fast-moving water in the gorge below. This place, where you catch glimpses of the river, is at the border between Tennessee and North Carolina, below Walters Dam, the current of the river manufactured by releases from the lake, the gorge blasted and reengineered to accommodate the interstate. You knew that this was the Pigeon River, that it descended from Sams Knob on Black Mountain, at an altitude of 6100 feet. But what you didn't know could fill an empty lake, and the dam of ignorance was about to crumble.

Where you've camped on the Pigeon, miles below the dam, near Hartford, Tennessee, you ask a river guide named Nate how to get onto the lake. "Why would you want to?" he says, squinting, a guide for sixteen years on the famous Green in Utah, the Rio Grande, the Nolichucky (his favorite), and the Ocoee; he talks with reverence about gradients and with awe about the braiding of currents, as if he were speaking of a lover's tresses. As he talks, it looks like he's seeing the rivers, not you, standing there beside him in the dark in front of the little store. You think he may have smoked some marijuana. You declare that you want to paddle up the dead lake to where the Pigeon resurrects itself, and you're telling this guy Nate you want to fish there. "I wouldn't eat anything out of there," he says. The paper mill's upriver in Canton. You remember what you read, about one river runner coming out at the mouth of the Pigeon, where you want to go, and saying he'd seen a giant ream of toilet paper, that he would never go back there again. The guide has no idea how to get on Waterville Lake, but he wants to know about it after your paddle; he wants a report. "Watch out for trot lines," he says.

Searching for a put-in, you pull off an exit called Harmon Den, and you see a nice little creek, mostly dry, mostly stagnant, that looks like it might make a fun paddle someday. This is no creek, you learn later; this is the Pigeon below the dam, the part that the power company has diverted away from the water that runs through the ten-mile-long tunnel from dam to powerhouse. No wonder you're confused. You ask at the North Carolina Visitor Center if there's a detailed map of the area. The woman who's stocking maps and brochures hands you a state map, no better than what you've got.

"What are you looking for?" she asks.

"I want to get onto Waterville Lake," you say.

"What for?"

"To fish."

You've brought a fly rod, a spinning reel and rod, and lures in a broken plastic tackle box, the contents of which you will spill three times during this excursion. Fishing, that's your cover, your alibi; it's almost normal, and you've read that people catch crappie, shellcracker, and blue gill from the lake. One guy said he'd eaten them for years and never got sick. Another said it was "the greatest lake in the world."

"It's mostly privately owned," says the Visitor Center lady, "but I've seen people fishing under the bridge at Fines Creek, and I think you might be able to get a boat down there. What you got? Canoe?"

"Kayak," you say, and you think you catch her shaking her head after you thank her and walk out into the heat. That night you sleep in a hammock next to the river dozens of miles below the dam, the powerhouse, and the paper mills, and you can smell the effluent, the distinct chemical aroma still there even after Champion closed down, even after the EPA clamped down on the company after Champion and the river no longer looked like coffee, here in this hip little town where most seem to make their living from the river one way or another.

The Pigeon rushes around an island and makes an eternal roar that's background to the intermittent groans of the interstate, the trucks gearing down on the climb through the pass. You watch the fireflies blink on and off against the dark rising bluff across the river and listen to the Cincinnati Reds lose a ballgame on your radio. Next morning at 7:30 you arrive at what you think is Fines Creek after a thirty-minute drive. You park on some fallen yellow tape that says "DO NOT ENTER: CRIME SCENE." Somebody pulls over in a white van and the passenger window comes down.

"You know sport?" says the driver.

"Do I know a sport?" you say. You walk toward them.

"You know *store*?"

You do not.

"We are looking for night crawlers," says the driver. You think the accent is Eastern European, perhaps Russian. You tell them that it is possible to dig night crawlers yourself, and they laugh. The Russians drive off in haste. Back to business. You're convinced that this is the worst put-in in the history of put-ins, and you've seen a lot of them: a steep fifty-yard drop down crumbled rocks slick with moisture, fringed with thickets of ankle-high poison ivy. You're hoping you don't disturb a copperhead. There's trash everywhere you look, bank-fishing byproducts, mostly notably a can of W-D 40 floating where you throw down your boat and huff to catch your breath after the descent with the fifty-pound kayak on your shoulder. But you try not to breathe too heavily because it smells like a river of piss, like an alley behind a bar full of serious and uninhibited beer drinkers. You're on the water, and you can't get away from the piss smell, no matter how hard you paddle through the stinking fog rising from the surface. Fines Creek is huge, and behind you it flashes white over rocks, forty or fifty feet wide. You'd think it was Fines Creek, right? On the Fines Creek exit on Fines Creek Road. You're only interested in paddling to the end of Fines Creek onto Waterville or Walters Lake, then turning left to find the mouth of the Pigeon, the egress, the vomitorium of the great river, which is taken advantage of to make paper for the books and magazines you like to read. Hypocrite.

An egret scolds you, crossing from side to side of the creek, which is widening into a fjordlike bay. Wood ducks do their false alarm flutter across your path to distract you from their young. Up a bay a waterfall bounces off a flat rock and splashes up not unlike the discharge of an industry. You could have put in there, had you known where it was by land. You see no humans and only two human habitations, high on the bluffs. The gathering sunlight whips the mists into tiny tornadoes, and the fjord widens. Trash cowers in coves and eddies. No sign of the Pigeon. You round a bend, and the dam appears, not more than a quarter mile ahead, a low gate with the big red-lettered signs warning you of the dangers the power company has gone to the trouble of creating. You're a bit dizzy with disorientation. You've paddled the length of the lake somehow, and you should have seen the Pigeon's mouth already. You see the mouth of Cataloochee Creek, coming out of the Smokies, up by the dam. The Pigeon should be somewhere, but it's not here. You turn around, fearful of the top of the dam so near, looking so fragile here in this gorge holding back so much water. You'd heard on the radio from a man who wrote a book that the weight of Lake Mead, formed by Hoover Dam, provoked an earthquake. You don't want them to let out the water yet, though you know they do every day. Will it sweep you toward the dam as it shoots from the penstocks at the bottom and propels Nate's rafting clients below? Will you topple over the brink and ride the river bareback to Foxfire Campground and your forlorn tent? No, that won't happen because this isn't a normal dam with tailwaters. It's got the tunnel that diverts the river to a power station miles downriver. You saw the dry stretch of the Pigeon at Harmon Creek, remember?

There may be good reasons no one else is on this lake in a boat. You're paddling back to the bridge to take out and go home, to give up in your quest to find the liminal zone of the Pigeon, and then you see the bubbles rise all around, like you'd seen other places where rivers come back to life. Carp, you think, gathering in a place where food is plentiful, gulping down whatever the renewed current is stirring. In the distance they jump, but not high. They are yellow-brown, impressive in girth, and belly flopping with gusto. On the bank you see the Russians who were seeking night crawlers. One of them stares at you and waves. How did they get there, you wonder; the interstate is one hundred yards above them, a steep climb. Something spins you around and makes you realize where you are. You had started out on the Pigeon, the place that had been your destination, not knowing it because the name of a road, an exit, had misled you.

When you're back at the bridge, you see it anew, the trashy place the river runner referred to, one of the dirtiest you've seen on the water, one of the hardest lakes to get to, one of the easiest river mouths to get to, though it's the naming that confused you, and the confounding and ignorance that the building of a dam creates in regards to the names and identities of rivers. Change the name of something, and you take possession of it. Your head is swirling. Nate the guide would be amused at your findings, but you decide to let him wonder, to let him imagine what you might have found on the lake that defies access, the lake that nobody wants to get on. He should be on the water by now, reading the current courtesy of the power company that guards the lake that funnels the water to the Pigeon where rafts full of customers hang on for dear life.

A year later, you're back on the Pigeon, this time below the dam, below the powerhouse with a group of three—father, mother, adult daughter—in a raft guided by a woman named Heather. It is raining lightly, but it's fairly warm this September day, and you all know you're going to get wet anyway. They told you so back at the outpost, where they issued helmets (brain buckets), life jackets, and bright-blue raincoats and pants. You are told how to act if you fall out; if you're a man, not to face downstream with your legs open, or something called "romancing the stone" will occur. Watch what you do with your paddle; don't give another passenger "summer teeth" (some are there; some are not.). They told you, in full disclosure, that the power company was not generating, so the river would be low. Having seen the Pigeon's liminal zone at Fines Creek Bridge the year before, you're on the lookout for suds, brown water, stench. You'd read that the conditions of the river had improved drastically in the last twenty years, that stricter regulations and improved industry practices had clarified the rapids you used to admire from the interstate at the lip of the gorge. Now you want to see up close, and you perch yourself on the edge of the big raft, wedge your feet tight between its floor and wall and do what Heather, in the stern, tells you to: "Forward! Back! Lean right! Lean left! Get up! Move to the other side!" She calls everyone "hon," even when she shouts, even though she's the youngest in the boat. "Bump!" she yells more than anything else because at this level, instead of rocking down class IV rapids, courtesy of Progress Energy if you'd come before Labor Day weekend, you're weaving your way through the exposed exoskeleton of the riverbed, narrow passages that keep Heather prying and drawing and yelling orders from the stern.

You go through a few passages backwards, and Heather smiles. The raft rams a rock, and she looks worried but smiles as the father, who has had heart trouble and wasn't supposed to be paddling, falls onto the floor and then rights himself. You want to help her more than you're able to. At a passage called "Lost Guide," Heather ferries diagonally across the forty-foot-wide river to reach a six-foot-wide passage between two Volkswagen van–sized boulders, a waterfall that descends six or eight feet. You hit one rock, and bounce backwards—stuck. A raft from a rival company slams you from behind, nowhere else for them to go. Water pours over the sides of the raft as Heather looks around, cranes her neck to assess the situation. You and the others scoot and pry at her orders. Not budging. The guide in the one behind breaks his paddle trying to pry you loose and has to take one of his passengers' paddles. Heather takes a deep breath and gets out of the raft, waist deep in the cold pouring water. She frees your raft, jumps back into the stern, and you plummet down the six-foot drop and bounce along on the waves. "I'm sorry, guys," she says. And the mother says, "Stop apologizing. It's us lard asses slowing you down!"

You're having trouble observing the water and the shoreline like you're supposed to be doing. But you've seen some softball-sized suds, although not many. The water is gray, a reflection of the weeping sky, and you do not notice any discoloration. You wouldn't describe it as clear, exactly, like you'd expect of a river coming out of the mountains, but you couldn't call it murky either. You smell rain, and you smell the familiar fishy scent of decay and death, nothing chemical. You take out at Harford, and the bus transports you and the others back to the outpost, where you take your nylon shirt off and let it dry on the hood of your car. An hour later you put it back on. If this is sweat you're smelling, something's wrong with you. It's not your sweat. It's a smell that reminds you of your father's clothes when he returned home from a swing shift at the chemical plant in Calvert City, Kentucky. You talk to river guides John Bowers and Iris Bahr-Winslow, and they tell you that your shirt stinks because of effluent from paper mill in Canton. Even a product called "Sink a Stink" will not take the stench from their clothes and life jackets. They believe the Pigeon should be as clean as it can be, and they are fighting against a renewal of the paper mill's permit that allows pollution to continue at what they consider an unacceptable rate. "They should make the river as clean as it is upstream of the paper mill," says Iris.

Heather keeps telling you to come back in the spring, when the water is higher, when the class IV rapids will make the run faster and smoother and wetter. You say you will come back. You've learned a lot from the Pigeon about abuse and resilience and damage and beauty. It deserves some attention. You promise Heather, you promise the river—that you will return. With time the shirt will lose its stench, you think, and maybe the river, with help and time, will cleanse itself.

NIGHT PADDLING

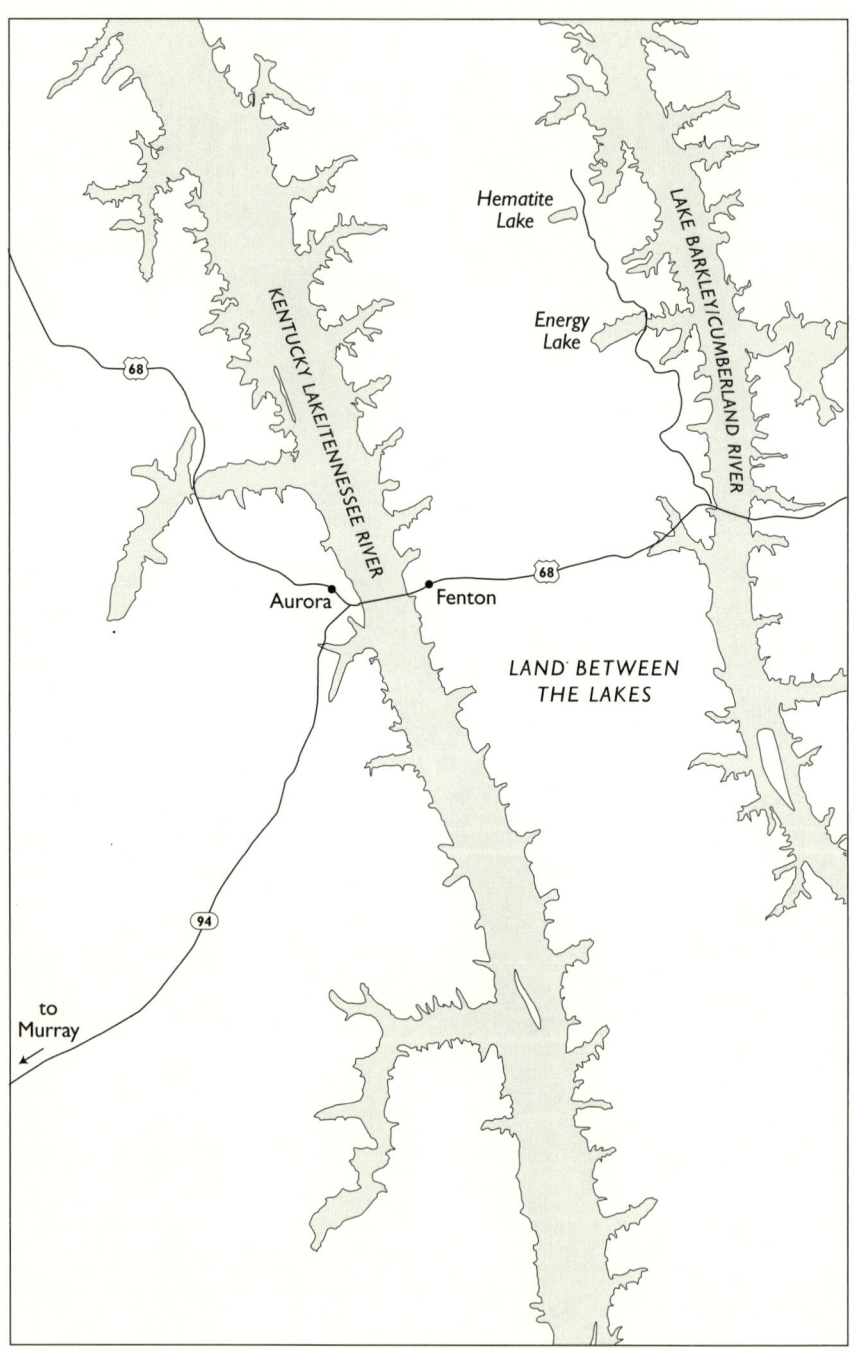

Land Between the Lakes

CHAPTER 14
HEMATITE

With the widening of the information superhighway and the ever-expanding role of technology, its presence apparently essential for minute-to-minute existence, we've become a culture that thinks it knows it all, knows how to do it all, and knows everything about every place possible to go and what to do there. There's an "app" to solve every problem, it seems, including running out of beer at a party where beautiful women dressed to kill crave a particular low-priced, low-calorie brand. I'm sure there's an app for exploration, for finding new places to go and new ways of looking at them, but I hope that technology will never wring the mystery out of nature, especially waterways that are alive with current. David Mazel, in his essay "Annie Dillard and the *Book of Job*," has noted the tension between the search for spirituality and mystery in nature and "global biosurveillance," the ability not only to understand the "biological whole" of nature but also to deconstruct it "layer by layer." This perspective is in some ways a threat to the sense of awe and mystery that God meant to impress upon Job, who was questioning God's power and knowledge after suffering a series of calamities he thought unjust, given his righteousness. With your limited perspective, what can you know possibly know about justice in the universe, God asks Job, and then gives him a series of questions he can't answer, such as the "birth season of the Ibex," many of which are easily obtained nowadays by anyone with an Internet connection.

    Like Job, I was formulating cosmic questions and seeking answers that went beyond science and empiricism. To an extent, all these excursions toward the liminal have been a form of night paddling, groping forward for something that's as abstract as fog, more difficult to grasp and examine than a rock or a fish or an antler. So it made sense to me, in the summer of 2011, to take the groping to its ultimate conclusion: to paddle blind and see where it would take me, to see what would arise from the rivers on the trip out and back. These would not be trips that commenced an hour or two before dawn. I'd done that at the end of my Tennessee River voyage, desperate to reach the end of the trip at the confluence of the Ohio. Also I had paddled through the dark on the Dolores River (see chapter 8) in Colorado when I set my alarm to the wrong time zone. Traveling toward imminent dawn gave these trips a sort of upward narrative arc, a Romantic optimism; they were voyages toward the light, as it were. These trips at night would have no such intrinsic arc. Commencing at some point after dark, they would end when I decided or when something decided for me. In a sense, these trips, all upstream toward a liminal zone, were sleepy-time attempts to use a cosmic dimmer switch to transform how my mind processed and made meaning from the landscape.

  I was spending parts of June and July in western Kentucky, where I grew up, near four major rivers and a slew of tributaries, creeks, swamps, and lakes. This was a good place to think about choices, immediate and long term. I was staying in what my family calls the old farmhouse, the house my grandfather Jack, a tobacco farmer, had built back in the 1920s, after he married my grandmother Otie. The farmhouse had been a place of respite and restoration for me several times during my life. When I got there in mid-June, it was surrounded by waist-high corn farmed by a cousin, Robert, and by the time I left in July, the corn was way above my head, tassled out, as my forebears would say, the rain and hot days making the ears seem to grow right before your eyes. A tall oak, in the process of dying, had shaded the yard since Jack had lived here. He worked most of his waking life, but at night he liked to lie on a blanket in the yard and listen to baseball on the radio. When it was not too hot, I'd leave the windows open and listen to the Cubs myself, in an attempt to connect with surly Jack, who was a Cardinals fan like most western Kentuckians.

  An unrepentant chewer of tobacco, Jack was not known to waste words, but he did teach me the rudiments of fishing in one of the two ponds that he dug and stocked, and then later inexplicably filled in to create more acreage. In true Trevathan style, he consulted no one, explained nothing, just did it (long before the shoe company's slogan). He was remote, larger than life, always in blue-striped overalls, sometimes without a shirt, and he smelled like the tobacco that he grew, the smoldering sawdust that coated the barn floor, and sweat. He used to walk beside his tractor in the field, shifting the gears, steering it when needed, as if he had a relationship with it similar to the mules he had used. The only time I remember him agitated and anxious was during his last days in the hospital, gazing out those windows that wouldn't open. He died a few days after that. Jack's outbuildings were still there at the farm, many of his and my father's tools lying around haphazardly in the wheat shed, patrolled by stray cats, snakes, and yellow jackets. The tractor, which still ran, sat in the garage. The tobacco barn and stable were long gone. My parents built their dream house—white painted brick, front porch across the entire facade—fifty yards from the farmhouse, and my mother had sold the dream home to another family, who now occupied it. No matter. The land buzzed with stories, with hurt and joy and hunger, and even though there weren't any rivers in the immediate vicinity, I could feel their influence. Jack and my father, Ben, who lived to work, it seemed to me, might not understand fully what I was doing and would rather see me making money or making *something* out of all this energy I was expending. Still, I think they'd be sympathetic to my desire to be outdoors.

Where to go kayaking at night? More important, how would I go about this? Should I use a headlamp? What misfortunes should I be ready for, paddling at night? Sneaky reptiles, rabid marsupials, angry predators? A swamp monster? If there's one thing I've learned over the years of camping and hiking outdoors in the dark, it was this: at night, anything is possible.

I decided to see if others were in the habit of paddling at night, what kind of folks they were, and what they had learned from doing it. Turns out that I wasn't the first paddler to think of this. No big surprise, though I was a bit surprised at all the rules and cautious advice the night paddlers seemed so intent upon. Take this. Don't take that. Do this. Don't do that. Everyone recommended taking a light of some sort, and then there was a big discussion on what kind of light would be best. Be sure to have an air horn, reflective tape, a flare gun. Nobody wanted to get run over by a cruise ship. Don't forget your PFD (lifejacket) and your map and your GPS. Most important of all, don't go someplace new. To be fair, this particular site I'm picking on was aimed at beginners, it seems, the night paddling focused on busy ocean embayments. Another site said to avoid paddling where there were lots of other boats. Another had meticulous advice on when to paddle according to a specific phase of the moon: this guy wanted me to buy a sun/moon chart. Now, there's nothing wrong with being informed, and nobody forced me to sit in a room and read this information. I sought it out. Some of it was useful, though I was struck by an overall sense of dread, something like the voice of a Boy Scout on Ritalin: be really prepared because it's, like, dark at night, and you can't see and others can't see you, so string up a wad of solar-powered Christmas lights so that others will be mindful of your presence and not cause you to disappear beneath the dark waters, glug, glug.

It was good to have all this stuff in the back of my mind, if not in the back of my car, as I set out an hour or so before midnight on the eve of Father's Day, a Saturday night. I chose a lake in the Land Between the Lakes (LBL)—the peninsula between the lower Tennessee and Cumberland Rivers (now Kentucky and Barkley Lakes)—where my father had taken me fishing and where I developed my love of the outdoors. It was a place freighted with sentiment. I hadn't been there in decades, and I wondered if a visit at night would rekindle those sentiments, or, with the liminal in my forethoughts, create a conflict between my present agenda and the past, the passage of time a soft-focus filter that would gloss over the ugly, mundane, and annoying.

Hematite Lake lies in the northern half of LBL, a national recreation area established in the early 1960s by the Kennedy administration. The government bought out the families living there and set about creating a commerce-free peninsula within the local lake culture where development of marinas, condos, and other retail would be prohibited. This would be a place owned by the public, a refuge for humans who wanted to reconnect with each other and to learn about the natural world and to an extent about life there in the nineteenth century. According to *Louisville Courier-Journal* reporter Joe Creason, writing in 1963 when JFK was proposing the idea to Congress, LBL would be "a remote, rustic close-to-Nature region in a part of the country where such spots that can be developed in such a way are disappearing rapidly." Creason also noted that the peninsula "drips with the kind of history and romance that appeals to tourists," specifically the vestiges of the iron ore industry—the mineral hematite was the main ingredient in the production—and the subsequent moonshine production in the early twentieth century, which was of such high quality that it was exported during Prohibition to cities as far away as Chicago and Detroit. Despite the informed optimism of Creason's article, the twentieth-century history of LBL was for some a bitter story of dispossession and manipulation by government.

In 1938 FDR had acquired 65,000 of the 170,000 acres of what was then known as the Land Between the Rivers; this acreage, bought from Daniel

Hillman and Sons, an iron manufacturer, became the Kentucky Woodlands Wildlife Refuge. In 1944 the closing of the gates on Kentucky Dam on the Tennessee River would flood thousands of acres on the western shoreline of LBL, and Barkley Dam, which would be finished in 1966, would flood another 5,000 acres. At the time of Kennedy's proposal, less than 3,000 people remained in what would become LBL. Understandably they did not want to give up their homes, regardless of the government's compensation. They didn't want to give up their lives there, to have to leave a place that had a worth beyond a banker's appraisal. This was a process enacted throughout the country, particularly in the Tennessee Valley. I understand their sorrow and anguish at being uprooted from their homes, which were located in such a beautiful and unique place, and I feel as much regret for the way this project came about as I did for the creation of Tellico Lake (see chapter 2). It must have been particularly infuriating here for the exiled and their descendants to have to see their homeland frequented by Boy Scouts and tree huggers and campers from all over the world, acting as if they owned it. At the same time, this place, as it exists now, seems as much or more like home to me as the town I grew up in, Murray, about a half-hour drive away. I went to a school camp trip at the Land Between the Lakes as a sixth grader and stayed in a cabin for a few days, my first extended period outdoors; there I saw how some respected the outdoors and even became better people for it, and others went kind of berserk, either brutalizing their surroundings or cringing in fear from all the things they thought would hurt them. I visited on my own as a teenager to fish and to hike and to boat, and on up into my adult years I'd go back when I had the chance. It calmed me to go there. My father would take me fishing to Honker, Energy, and Hematite Lakes, bodies of water small enough that you could see the boundaries all around you that had been formed by small-scale earthen dams. Hematite was built in the late thirties as a "federal work relief project," according to Frank E. Smith in his book on LBL. TVA constructed Honker, Energy, and Bards Lakes to create inland lakes in an area where the high dams had flooded out many of the natural springs. The small lakes were intended to add to the aesthetics of LBL, creating wetlands rich with wildlife, including eagles, where without the small earthen dams, the drawing down of Barkley and Kentucky Lakes would have exposed bare mudflats. The smaller lakes, in addition, would "provide educational and recreational opportunities not available on the broad expanses and sometimes rough water of the larger lakes," according to Smith. "They provide a safe haven for small boats and canoes." Amen.

An early fishing memory of mine has its source at Hematite. My father and I fished Long Creek below the little spillway and farther down, where it was shady and quiet and narrow, isolated and mysterious like the liminal zones I now sought. I remember pulling beautiful fish from that muddy water, and I was young enough and new enough to fishing that it seemed like a miracle when something took the bait and zigged and zagged, and you could pull it from the water and hold the thing in your hand, then try to calm it and get the hook out and let it go again, or if it were big enough, keep it and consume it later. In that way it became a part of you, this fish and the creek from which it emerged.

Near Hematite's spillway, where we parked, there was a large pen that displayed a herd of deer, a couple of acres of woods where the captives moved silently in deep shade and allowed you glimpses of them. It's a bit odd to think of the deer pen now, when it's so commonplace to see them running wild all over LBL. Up the road a ways was (and still is) a nature center with stuffed and live animals for educational purposes, including the lethal copperhead and water moccasin. You could hike around Hematite Lake, 2.2 miles, and at the top of the lake, where the creek came in, wooden foot bridges took you through a wetlands where beaver added their engineering talent at pooling the water. I took nephews Sam and John and niece Mady there when they were six or eight or so, and marched them around the entire lake. I remember getting worried about them about a mile in when it didn't matter if you turned back or kept going because it was the same distance back to the car. As a sort of reward for surviving the long hike, I let Sam sit in my lap and steer my old car for a mile or two. Now, fifteen years later, they still talk about it.

Upon pulling onto the empty gravel lot at Hematite in the summer of 2011, I got out of the car, determined to perambulate without my headlamp. I found myself groping awkwardly as the interior light from my car faded away, and the once-familiar landscape, laden with sentiment and important personal symbolism, now seemed alien and indifferent to me and my desire to become reacquainted at this strange time of day. I stumbled over roots and rocks. Stomped through a large puddle. Resigned to having to use the artificial light until the moon rose, I began looking for a suitable spot to put in, for there was no boat ramp here that I knew about. I'd never seen a boat on this lake, come to think of it. I trained my tiny headlamp on the spillway: it was not spilling, its concrete apron bare or holding shallow pools of fragrant water. There wasn't enough water to put in at the steps that went across the top of the spillway. As dams go, this one was about the least intimidating I'd ever seen, though as a boy, never really having experienced the high dams on Kentucky and Barkley Lakes, it never would have occurred to me to compare Hematite Dam with a high dam on a major river. And if you were a bitter former resident of LBL or an ancestor of one of them, I doubt you would direct your rage at Hematite Dam but at the larger dams which flooded out many (but not all) of the farms, communities, and cemeteries.

As I found out later from Les, who worked at the nature center, the lake itself was excavated as a result of the nineteenth-century iron-ore industry. The mining of hematite for the process of producing iron created the lake bed, and then the dam was built later to stabilize the lake and help the wildlife flourish in the wetlands there. The spillway was the type known as "uncontrolled" or "free," which, according to engineer Alfred R. Golze, "automatically releases water whenever the reservoir water surface rises above the crest level." Golze called this the simplest, most maintenance-free type of spillway. To me, Hematite, especially compared to the high dams, presented a seemingly benign, accessible, and underwhelming human structure, particularly with the square concrete stepping stones you could walk on to cross the top of it.

Les told me that back in May, when the Mississippi and Ohio Rivers were flooding, the water had risen eight feet above the road that was below the spillway. In fact, after an early spring of killer tornadoes and torrential rains, including a freakishly potent hail storm in Knoxville, the Mississippi had swollen to a height not reached since the great flood of 1927. By June the rivers and lakes had receded to normal levels. Back in East Tennessee, it was very dry, no rain for more than three weeks, and in western Kentucky there had been plentiful rains and storms. Meanwhile parts of the Southwest were parched in a record drought and with searing heat. To say the least, 2011 at that point had been a year of extremes in weather. And when weather is extreme, no attempt to control nature—whether it's a levee or a dam—is guaranteed. Even though Hematite covered only ninety acres, in 1998, after heavy rains, its dam failed and drained the lake. TVA repaired the dam early in the summer to accommodate boaters, fishers, and hikers whose numbers peaked that time of year. This was one dam, I have to say, that I was glad to see repaired.

One thing different about night paddling at Hematite versus other lakes was the complete absence of artificial lights. There would be no barges, no cabin cruisers, no porch lights or radio or cell phone towers on land, not even much air traffic above. Because of that expectation, I was dismayed to discover bank fishermen down the way; they had the headlights of their truck on intermittently, and they were playing their radio loudly. I kept my distance and used my light as little as possible out of respect for their privacy and their right to experience night nature in their own way. They probably expected to be alone as well, and to their credit, as soon as they became aware of me, they turned off the radio and limited their headlight use. I doubt this kind of enlightened and accommodating response would be as likely to happen outside LBL, which, I think, cultivated considerate behavior among its visitors.

Down a mown path I carried my boat and set it in the weeds growing up through the water. The moon was just beginning to rise as I floated out into the dark, calm water, and the guys on the bank cast oblong bobbers that glowed red, like something you'd wave around at a rock concert. I was doing this in honor of my dad, who introduced me to places like this, but who would also wonder what I was thinking, boating on a lake at night. *And not even fishing.* (He and my grandfather ridiculed a neighbor who plowed his fields at night.) One thing I did not have to worry about here: other boaters. No motorboats were allowed on Hematite. Ever. And nobody like me, in a boat without a motor, was motivated to paddle this particular lake tonight. I stayed in the middle, away from the guys' fishing area, and paddled toward the back, where the beaver dam used to be, thinking that the creek's entryway would be in that area. The insects laid low, and the air was still and cool, at least for summertime. As the moon began to rise, things emerged in profile, then in greater relief, a subtle dawning: the shadows of trees lined the lake, a jagged dark horizon on the surface that I paddled toward but never seemed to reach. There were the tiny splashes of fish tails, and some kind of pale matter floating in small clumps on the surface. The way things were emerging from the darkness, the way the darkness was beginning to fade in the moonlight, made everything seem alive in a way I hadn't considered before. The *gotcha!* feeling we often expect in darkness was giving way to something calming. I still felt a bit like an intruder in this night world, not of it, like everything else. I was the Other, and those who belonged were tolerating me. So far.

Paddling easily toward back of the lake, watching the clouds pass over the moon, I was glad I'd made the effort to drive, with one headlight out, the forty-five minutes from Murray. It wasn't so bad, I thought, not to be completely alone out here. The fishermen had turned the radio off, and they were quiet. I wasn't certain how far away they were. After I'd paddled almost an hour, big shapes began to emerge, aquatic bushes and clumps of lilies that protruded from the surface and reflected a shocking green when I shone a light on them. I had to raise my rudder when it began to drag bottom. Little splashes plipped all around, and from the mid-distant came the croaking of bullfrogs. Nearer, in the shallows around me, smaller frogs made a clacking sound like castanets. A heron rustled its wings and squawked as it flew away, disturbed from its first sleep. Once I got into the maze of plants and the woods at the back of the lake, I felt more and more like an intruder, as if something might arise from the swampy waters and give me a sign to turn back. I poked around for a passage but came to dead ends in my pursuit of the creek that fed the lake. Then came my sign: lightning repeatedly flashed silently to the north, a strobelike cosmic signal. A gentle headwind arose on the way back, as if to challenge me to a race with the coming storm. I paddled toward the light of the fishermen's pickup truck, about a mile away, and the lightning increased in frequency. A distant rumbling emerged once, twice. I liked paddling at night, I decided, but the prospect of bad weather seemed even more daunting than it would be in the day. As the moon rose higher, I got more and more adapted to night vision, but now clouds were shrouding it and thickening to the point that they would blank it out soon.

I approached the fishermen, who were near my takeout. They were Texans, one from Lubbock, the other from Houston. Young soldiers taking a break from their training at Fort Campbell, they had been fishing here for the past two days. The Lubbockian said he'd caught some bass, including a four-pounder, the day before, but tonight they'd caught only a couple of catfish. They clearly loved the little lake and asked me if I'd caught any. I said I was only taking pictures. "So that's what that flash was," he said. I pointed out that I'd also seen lightning to the north, which they hadn't noticed, the ridge rising at their backs blocking the sky. "Be safe," said one army kid as I paddled off, "and don't turn over in that thing."

I'm conflicted about these small spillways, about larger dams and the forces behind this recreation area for the public, and about the high cost to the families who lived here. Clearly it was made possible by the government's power to build large structures and to have and to carry out a unified vision. Even though the goals included environmental education and even though commercialism has been excluded for the most part, much of LBL is artificial in the strictest sense, particularly the little lakes I love. Should it have been done differently? If so, how? Why couldn't we have had LBL without the big dams, and could we have allowed the few families to stay there on the condition that others could not move in and crowd the wildlife?

I went back to Hematite in the daylight to try to resolve these questions and to press farther into the back of the lake through the mazes to the source of Hematite's flow. It was a glorious morning, coming after we'd had five inches of rain in Murray over two days, the last day a wild party in the sky that started around 2 A.M. and lasted until 4, the most lightning and thunder I've ever heard in one stretch, nonstop, for two hours. On the morning of my Hematite reprise, the world was fresh and clean and clear as it can only be after a deluge like that. It was almost 7:00 before I was able to get onto the water, the sun fairly high above the horizon. But the atmosphere was clear of haze, it was cool, I was back on the Hematite, and this time, on a Wednesday morning, I had it completely to myself. The first thing that struck me about the daytime lake was that everything looked smaller than it did at night. I paddled to the end of the lake in what seemed like an instant. The bush that had emerged from the nighttime, its tendrils ominously reaching for the black sky, seemed innocuous, one of many pesky obstacles I had to weave through to get to the transitional zone. These visual revelations demystified the lake. It did not seem bottomless as it did on the night paddle. Quite the contrary. Not only could I see the bottom half a foot below me, I could also see that the bottom was covered with some kind of thick-stemmed scummy weed. It was everywhere. I would rather it wasn't there, but it was, and I remembered that, yes, this was part of Hematite, that it was shallow and full of weeds, and that's probably why my father preferred fishing below the spillway and not on the lake. I reconciled myself to the weeds and paddled onward, wondering how I missed the big wooden wildlife pier and platform the other night, off to my left.

Perspective was another thing missing from the night paddle, that sense of knowing the entire tableau of which one is a part. This time I forged through the sea of bushes, into the forest of water lilies, straight through the passage that led to the footbridge I remembered. Some of the water lilies were blooming yellow, and bright green pads the size of catcher's mitts covered the entire surface of the narrowing channel except for the path I was taking toward the source of the water, Long Creek, somewhere back at the end of this maze. Silver-dollar-sized droplets of water stood on these pads and trembled as

I disturbed the water ever so gently with the tip of my paddle. Toward the footbridge I floated, as quiet as I could be, though every sound I made seemed like a Mongolian pounding a war gong. The floating pads and delicate lilies I tried to avoid in feeling my way forward seemed fragile and beautiful like something someone might plant in a pond to look at, not to swim or boat through. As I approached the bridge, which had iron support bars built beneath it, I saw major movement on the gravel bar beyond it, maybe twenty feet from me. Two deer straightened up from their sipping to stare at me, then edged away and disappeared as I fumbled with my camera. Onward I paddled to one, two, three dead ends, dark places where the water was black and clear, similar to the blackwater swamps of the Edisto River (see chapter 10), except there were no alligators. Frogs grunted but not as loudly as at night; one of them started up with the castanets again. Red-winged blackbirds, indigo buntings, and goldfinches chirped and flitted about. In the peace at the back of that lake, where the forest closed in and the channel narrowed, I explored every nook and cranny, looked up high in the trees and low into surface that reflected the tiny yellow buds and the fanlike leaves. When I found the trickle of current, it was unassuming, and it disappeared upstream into a thicket of wetlands too narrow to navigate. Hematite was subtle and secretive even in the daylight, and I was glad, in a way, that its liminal zone was too small and shallow for me to navigate.

At the back of the maze near the trickling of Long Creek, the daytime experience of paddling Hematite approximated the peaceful and slightly eerie feeling of the nighttime trip. I had more night paddling to do, and I would continue to probe these kinds of places in LBL, familiar to me in some ways, but also remote, made alien by the lapse of time and the absence of light.

CHAPTER 15

ENERGY

I got to Energy at 10:30, just as the waxing crescent moon was about to drop below the horizon of the swamp I was heading for. It blazed orange on its slow descent, as if aflame. The stars shone in great numbers, crisp in their celestial grandeur. Flashes troubled the horizon all around, and thunderous rumbling signaled that the last pseudo-weaponry of July 4, 2011, was being expended in a frenzied patriotic orgasm. A big splash, like a bowling ball dropped into the water, made me wonder what was out there, what I'd see and hear tonight. I considered getting into the car and driving back to the farmhouse to go to bed, yet something drove me forward, a curiosity about the night life of the Crooked Creek swamp that I'd already invaded the day before. I anointed myself in bug spray and attached my headlamp, vowing not to use it unless I really had to, on the theory that my eyes would adjust to the moonless night. The water seemed thicker at night, my paddle strokes quieter, as I tried to minimize extraneous movements. I cinched up my uncomfortable life vest and advanced upon the darkness.

    Having paddled Energy in the daytime, I knew that finding its liminal zone at night would be an ambitious goal, near impossible, even though the entryway was marked by two trees that had fallen together, making a sort of cross, or X. After that, I entered a full-scale natural labyrinth, challenging even in the daytime, the hedges and grass islands forming multiple passages. It seemed as if I were entering the rooms to a large house that were separated by narrow hallways. Often, I'd paddle until the walls of vegetation closed in and the passageway transformed itself from creek to ditch to watery meadow. Sometimes the rising of the muddy bottom would stop me. Finally, I followed a sort of zigzag pattern and scooted over a low spot to an island where beaver had built a house of sticks and mud. I could hear some work going on within, like someone patting the mud with the flat of his hand. Just beyond the house a small raccoon sat staring at me. He retreated a few feet and peeked around the stick house. Just a bit beyond it, a big crashing noise and a series of splashes told me I'd flushed a deer, and as I continued following this course, the deer huffed repeatedly in that shrill voice that sounds like a mix of anger, fear, and exasperation, an expulsion of air that comes from deep in the throat. I disturbed the same heron a couple of times, and its cries seemed supersonic in the quiet, closed-in area. High above, a woodpecker launched into a project with such force it seemed as if it were swinging a hatchet against the trunk of the sycamore. I expected the tree to fall at any moment.

Now, at night, I paddled near the bank below the campground, still a mile or so from the crossed trees. Up on the hillside three shapes sat beside a campfire talking, two lower voices—young men—and one higher voice that I imagined belonged to a young, impressionable boy. I heard the word *scary* a few times, but I could make out little else. The tone of the conversation was sober but intense.

"Is that a boat down there?" the higher voice asked the others.

"Is there someone down there?" asked one of the deeper voices.

"Just me," I said. "Hope I didn't scare you."

I wondered what I looked like, what I sounded like from up there in the glow of the campfire, on solid ground. There were a few security lights in the vicinity, but I felt much more obscured than they, and if this were the Civil War and I part of an attacking amphibious force, I don't see how they could have prepared for my attack. On the other hand, a warship would probably not be bright yellow. As it was, it probably gave them something to talk about for a while: the crazy guy in the kayak, going where, doing what? Alone . . . at night.

Energy had none of the sentimental trappings of Hematite, none of the personal history. The dam, an earthen structure reinforced by limestone riprap like the spillway at Hematite, was a half-mile long with a road running across it wide enough for fishermen to park their cars on the side and toss out a line from behind the wheel if they wanted. In the daylight Energy was not as interesting as Hematite. Its scale made it more ordinary, its features blander, less pronounced, than Hematite's weed beds and the steps across the top of the spillway. At 370 acres, Energy is a "subimpoundment" of Barkley Lake, constructed to provide "stable water levels for water-based recreation at . . . Energy Lake Campground . . . and the Youth Station resident center."

I paddled past the last light from the campground bluff—about a mile in—and advanced with less conviction toward the shadows at the back of the bay, another half mile farther. I could see the outline of the trees on the horizon and gauge from that how far from the bank I was, but I had little sense of where to enter the swamp or where to find the crossed trees that marked the passageway to the inner sanctum of the deer, raccoon, beaver and who knew what else. What was most visible loomed above me: a night washed clear by another good rain, stars blinking and shooting across the sky, an aircraft or two. The fireworks had ceased by the time I reached what I thought was the perimeter of the slough. I stayed left and thought I was entering the swamp but found myself almost colliding with the back of a shallow cove that I remembered from the daytime voyage. Seeing the bank where I didn't expect it was like the sensation of running into a wall at night as a guest in an unfamiliar room. I felt my way out of the cove and turned west toward the lake's terminus at the next opening. I could feel more than see the banks closing in, though I could make out the shapes of the trees growing closer as I paddled forward. What I could not see at all, no matter how hard I strained, were the bushes and grasses that began to . . . well, not exactly appear, but come to be as I went forward into the heart of this darkness.

In the daytime maze earlier a doe had appeared on one of the marshy islands; just her head had been showing behind a stand of weeds about fifteen feet away from me. She stared at me, her ears up, comically large, I thought, for the size of her head. She moved toward me a step or two, not taking her eyes off me. After a minute she stomped a couple of times, as if she were trying to see how I would react to intimidation. I wondered if she were trying to scare me off, to flush me like a person might try to shoo a cat or a possum. I'd read of deer attacking humans, particularly during fawning season when protecting their young. But she seemed to have a friendly demeanor, and I flattered myself that she was pursuing me, that the little stomps were messages of some sort, a kind of hello in deer code. After three minutes, she fled, splashing and making a huge ruckus in her flight.

Now, groping blindly toward the slough, I gave up waiting for my eyes to adjust and turned on my headlamp. The feeble outlines of the bank disappeared into a soup of swirling fog, a whiteout of the landscape as complete as a well-executed special effect in a horror movie, the kind that are so often made in places like this. How many, beginning with *Swamp Thing*, have there been? If not a swamp, then the woods. If not the woods, then a large body of water. Did the serial killer Jason's hand not arise from the lake at the summer camp in *Friday the 13th*? Was there not a horror movie called *The Fog*? This fog, my fog, swirled, as if alive, into levitating shapes that formed faces and monsters. Perhaps I would be devoured in some horrible way and never be heard from again. Had I not seen the designation "lethal fog" one morning on weather.com? Where I was—and it just occurred to me the creepiness of the name "Energy"—combined all the elements of scariness. It remained only for me to find out how far I could tolerate it and whether the merely scary would escalate into a *situation,* a tangible threat. Would I scare myself to death in a benign and peaceful landscape? The swamp was at full-volume stereo, filling up every available space, every direction, but the night shift had taken over for the daylight animals. Hoo-hooo-hoo-hooo-hooo, called the owls from one side to the other. Frogs were belching low and rude, so loud it made me think they were big enough to swallow me, boat and all. The fast clapping of the castanet frogs made my ears flinch. From somewhere behind me, it sounded as if something had waded out from shore, the scaly two-legged lizard with the big teeth, no doubt, pissed off that I had the temerity to paddle my stupid yellow boat into his private zone of evil. Perhaps it was the doe, come to kick over my boat and stomp me into submission. Onward I went into the fog, unable to resist, drawn inward toward the great unknown.

At one point in the winding daytime voyage, I'd heard the cheerful, revelatory sound of a waterfall coming from around a tight bend in the stagnant slough. This seemed a wondrous thing in western Kentucky, an actual waterfall at a transitional zone. This I'd never seen, east or west. I got closer and closer to the sound, kept bending with the aptly named Crooked Creek and paddling slowly, softly. What I saw made me laugh. A steel pipe was gushing water from the bank, and at first I was thinking this was some kind of overflow for runoff, like a storm outlet in a city, a big joke on me after I had been imagining a pristine waterfall at least a few feet high. I got out of the boat on a small gravel bar, a rare sight in this swamp of muck and grass. The pipe ran under a gravel road from a spring, a small bowl-like pool at the base of a rock wall, about big enough to bathe (or baptize) a baby in. The water was ice cold, a contrast to the tepid lake, and below the pipe, a growing pool of clarity merged with the cloudy swamp soup. The range of my emotions had gone from expectant/excited to disappointed and back again to being delighted by the fresh spring and the fact that the creation of Energy Lake had probably prevented this spring from being flooded by Barkley.

Now, still at the perimeter of the maze, worlds away from the cheerful waterfall, the passageway narrowed, the fog thinned out a bit, and I navigated by the beam of the headlamp through the path that the grasses demarcated, went forward until I could advance no farther and sat there feeling more alone than I could remember, remotely at peace because, if nothing else, I'd reached a sort of goal in pursuing one path to its terminus, never mind that it wasn't the endless zigzag I'd taken in the daytime. I had to back out of there it was so narrow, a feeling that was like stepping into the air onto something you thought was there but wasn't. It alarmed me to run into a bush, and at the first opportunity I got the boat turned around by doing a three- , four- , or five-point turn. Back to the open water I headed, perhaps just a bit more confident, though my tailbone was beginning to ache, my legs to go numb from the way I'd held them still and rigid against the footrests.

Instead of heading back down the open lake, I continued parallel to the threshold of the swamp, curious about the location of the crossed trees that marked the entryway I'd taken the day before. Did this cross—more of an X really—signify a "do not enter" tonight? I wanted to find it to get my bearings. Despite having come here and spending hours in the maze, I now had no sense of the landscape's contours or of my location. I paddled into a large opening that I followed for fifty yards it seemed, the trees skeletal as they reached up through the darkness, even the ones with leaves on them. My lamp beam was no good here. The fog swallowed it up. I took a stab at finding an opening in the jagged dead-tree horizon. Turned on my light at what seemed a familiar spot, and there they were, as if the fog had parted to show me the crossed trees, standing in swirling wisps of vapor. Softly I paddled past them, wondering how far I might go. The first room past the threshold was large, closed off in a way that the previous passage was not. It was the antechamber, the foyer of the mansion where reptiles ruled. A heron squawked from somewhere deeper. I hesitated at the first sharp bend to the left, popped my flash a few times at nothing in particular. There was nothing to focus on in the foggy dark, so I was shooting blind, hoping that something interesting would wander into my shot.

I had wandered into what is known as a spiritscape, I think, fraught with numina, "spirits, deities, or a divinity inhabiting sacred sites which have certain powers that are commonly described as supernatural or magical," according to the glossary in the book *Sacred Natural Sites: Conserving Nature and Culture*. Though, as far as I knew, Crooked Creek slough was not listed on an official roster of sacred sites, as say, Machu Picchu in Peru or Stonehenge in England, there was a vibe out here that suggested, no *insisted* on, something beyond the "natural." I'd felt this before on the water but never so strongly as I did here. What I sensed had no particular identity. It was neither good nor evil, and I can't say that it was benign or passive, either. By this time I wasn't exactly relaxed, but I was alert in a different way, not jumpy or as reactive as before. I was beginning to get used to being here and to note things outside the realm of my psyche. At least I thought so.

What turned me around, what terminated my mission was not a frightful sight emerging from the chaos of the swamp to grab my soul in its gnarly grip, not a howl or a crashing through brittle twigs, not an explosive splash that signaled the immanent capsize of my fragile shelter from the void. It was a warning, for certain, both subtle and insistent, unseen but concrete in an olfactory sense. The Horror! The Horror! My flashbulb, my incursion, had alarmed a skunk—because, yes, as the poet Robert Lowell would say, it was the Skunk Hour—and I backed on out of there, heeding the warning respectfully, for the skunk rules, and I did not want to incur a direct hit. I'd seen what a direct hit looked like on my dog Jasper, what it took to clean him up, and I was looking forward to a return to civilization. A direct hit would mark the wilderness's claim on me for an extended period. For fifty yards of paddling, the skunk's warning hung in the air. I began to wonder if I had set off his alarm or if it were something that had nothing to do with me and my pathetic lights. I paddled toward the twinkling fires at the campground. Beyond the ridge a bigger beam searched the horizon, a barge passing under the bridge over Barkley Lake at a place where the channel twisted so torturously, it was called Devil's Elbow.

Below the camp with the talkers, I paused, hoping they'd say hello and come down to the bank and talk. I think that they saw me and someone made a brief comment, but the conversation was in full force, and what I'd thought was an impressionable boy, a Cub Scout, was in fact a young woman, I could tell by the tone of her voice—more assured, her comments fully developed—even though I could not pick out more than a phrase or two from below, something about "a few bad apples" at one point. I sat there and listened and wondered if they could see me down here eavesdropping. I hesitated to say hello from the water, for fear that I would startle them and they would say something like "What the hell do you want?" So I waited down there for a few minutes for them to say something friendly, and when they didn't, I paddled back to the ramp, glad to be back on the land.

What I felt that night—so different from the night on Hematite—operated on a deeper level, I think, from that mix of fear and fascination that makes a journey unforgettable and surreal. You wonder about the sounds, the shapes in the fog, and the brain is not calm, at rest, as I imagine it to be among those who meditate, but it is squirming, as Jim Morrison might say, reactive to stimuli, though at the same time those images, those feelings, created vivid moments as beautiful in their mystery as anything I'd seen on these trips. And this time I wasn't sure I'd be back. Though maybe, if under a full moon.

COMPANY

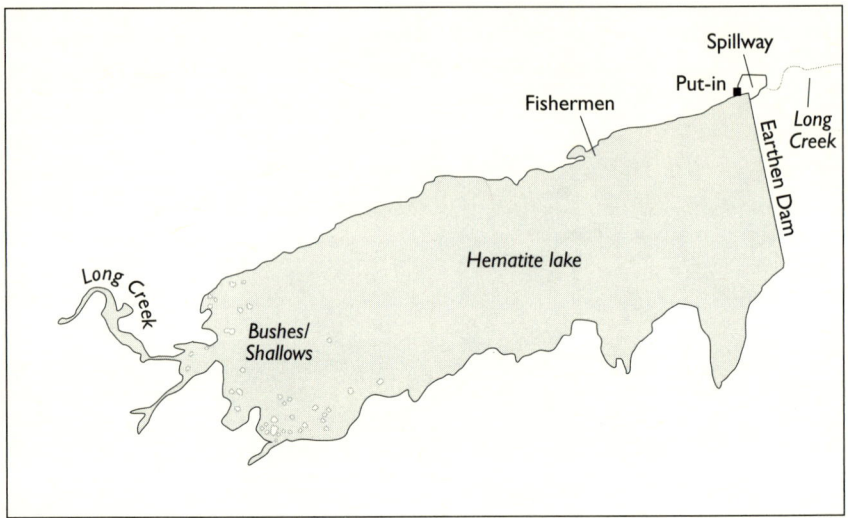

Hematite Lake

CHAPTER 16

WITH LIBBY ON HEMATITE

I started this project in solitude, convinced that it would be best to forge my way upstream as far as I wanted without having to consult with or worry about anyone but my own fool self. After a while, around four years, to be more precise, and particularly after the forlorn profundity of the night-paddling project, I felt that it was time to invite someone to come along on one of my liminal errands. It would be none other than Libby, my nineteen-year-old niece, on leave from college at Centre, a small liberal arts school in central Kentucky, where she was double majoring in studio art and environmental studies. Libby was just beginning to explore her interest in the outdoors, having dabbled in hiking, biking, boating, and rock climbing. She had recently survived Bonaroo, the Tennessee version of Woodstock, a kind of camping trip with hundreds of thousands of your closest friends. When I asked her, she acknowledged the presence of one of my contemporaries, Neil Young, who sang "Old Man" at the big festival, as I had hoped, though there was something wrong with his sound system and people complained that they couldn't hear. Her favorite was something called "My Morning Jacket," which reminded me more of a private joke than a catchy name for a band. "I grow old," as T. S. Eliot's Prufrock says, and it was time to infuse my journeys with the exuberance of youth. I'd told Libby about my night paddles on Hematite and Energy, emphasizing the "creepiness" of being on the water in the dark. This prompted her to ask if we could go at night. I flatly said no. *She* could go at night, but I, who would be answerable to my brother and sister-in-law, would not be the one to take her. We would go in the daytime, sort of, at five in the morning. I paused for her to challenge this ungodly time. No response. This pickup time, I lectured, was necessary, in order for us to be on the water at the best time of the day, just before the sun rose. We'd see more animals, there would be no wind, and we'd have the lake to ourselves. She nodded, as if this made sense to her.

There were three areas of concern on this little trip:

1. That I return Libby home unharmed and untraumatized;
2. That she not be bored;
3. That something interesting happen, some obstacle arise, something unusual or challenging to make the trip memorable, to create a story to tell.

If the opportunity arose, I was willing to be the one to sacrifice my dignity so that Libby would have an indelible narrative memory to tell her friends around a campfire.

All manner of obstacles can arise on a kayak trip, the most prominent of which is falling out of the boat. At Hematite, a small lake, that wouldn't be disastrous, though it would be a bit icky, requiring you to wade in the scummy weeds to the shore to get back into the boat. There were snakes, of course. The first thing my wife's father asked me when I told him I'd taken Libby for a kayak paddle on Hematite was, "Did you see any water moccasins?" Seeing a water moccasin wouldn't be so bad; getting bitten by one would give us a story that might not have an amusing ending.

In my brother's driveway at five, I had to call Libby to wake her up. It took her about ten minutes to come outside. That she was awake and talking alertly at 5:15 impressed me. There were other things I was learning about Libby as well. Though she seemed to have a lot of friends, she didn't mind doing things alone. While I was staying at the family farmhouse in western Kentucky, she regularly came out for solitary bike rides. This trait, the willingness to interact with nature alone, without company, seemed to have been passed down in the family genes, possibly from Libby's paternal and maternal branches: through my brother, a Trevathan, of the swarthy seaside-dwelling Cornish race; and through my sister-in-law, of Norwegian and Viking descent by way of Minnesota's North Country. This ensured a combination of toughness, courage, and thirst for adventure. It also helped that the rather slight build of both sides of her family—Trevathan and Engh—resulted in a physique ideal for kayaking. Though of Viking descent, Libby is quiet and gentle in manner, a kind of a twenty-first-century hippie child. She assured me that she'd been in a kayak before, though she did not elaborate upon that experience.

I put Libby's boat—borrowed from my sister's friend Mimi—in the water, gave her a paddle and life jacket, and pushed her off, thinking that she'd drift there for a moment to await my expert instruction. Was I not, after all, the guru of kayaking, the one in charge here? By the time I had my boat in the water, she'd zigged and zagged a half mile. Mimi's was a fast boat, lighter than mine, though without a rudder, it was a little more difficult to steer. With some difficulty, I caught up with Libby, who sat in the borrowed kayak as if she'd been born there. It was now my job to explain the difficult process of finding the liminal zone. "Have you read *Heart of Darkness*?" I asked her.

"Once in high school and once in college," she said. "I've written three papers on it."

"This thing I'm doing, it's like *Heart of Darkness*, going upriver to the earliest beginnings of humanity," I said, condensing it a bit. "Wait a minute. You wrote three papers on the thing?"

"One in high school, two for the college class."

The Horror.

I told Libby that it was up to her to choose the path to the liminal zone. I didn't let on that I'd already found it, the little trickle beyond the gravel bar where I had surprised the deer drinking water.

"You want me to choose?" she said, as if suspicious of my intentions.

"Choose our path to the source of the lake," I said with drama.

Instead of blundering too far right or left, as I expected, she wove a beautiful path through the bushes, between the lilies, ducking gracefully under the bridge and nosing through the narrow passageway between the gravel bar and the clump of lilies to where you could hear the trickling of water that was Long Creek. It shouldn't have been this easy for a novice. It seemed to me, though I didn't say so, that she might be better suited to these quests than me. After I had to scoot across the mud in a shallow place where I dragged bottom, I asked Libby how much she weighed.

"102," she said.

To create some more interest, I pretended that the liminal zone she found wasn't really the actual Hematite liminal zone, and we searched some more in the small slough through the passages in the water lily jungle. It could be that I had been mistaken, that there was another passageway to the watery cascade of Long Creek. Truly, Conrad's steamboat pilot Marlow would have been appalled by the tortured course of our meanderings that morning, and by eight o'clock or so, I was exhausted and hungry, and we broke out of the lilies for the open water. In the course of paddling the perimeter of the lake, I said at one point, "I could go to sleep right now." Libby agreed that she could, too. It had been a peaceful, if uneventful paddle, the liminal zone discovered after a half hour, the only animal sightings the wood ducks we kept panicking, the heron who kept flying away from us, and the loud, invisible frogs. I spotted a big flock of geese up in a cove, peacefully minding their business. I thought I might get Libby to paddle toward them and get a shot that would include her and the geese as they flew up into the air. In short I was attempting to use my niece as a sort of geese flusher, much as one would use a dog on land. She didn't seem to mind, but when she nosed the kayak gently up into the back of the cove, the geese merely walked onto the shore and across a little peninsula to another section of the cove. I paddled up to that section, and the geese walked back across the cove toward Libby. Either they were unable to fly or just stubborn, so I took the photo as they stood on the path like a crowd of goose commuters.

I explained to Libby how hematite, the mineral, was mined to make iron ore, and I described the fishing trips her grandfather, my father, had taken me on. I told her about hiking around the lake with her siblings fifteen years earlier and letting her brother Sam steer my car when he was six or seven. When we landed at the ramp, she said, "That was pretty. . . ." I thought at first "pretty" was an adjective, and I waited for the noun to follow, as in "pretty cool" or "pretty amazing." But that was it: pretty. Okay. That would do. It would have to.

The water had been so low that the liminal zone hadn't announced itself as emphatically as I would have liked, but after all it was the search that mattered the most, and it was all worth it to realize that my niece would rise at five in the morning, accompany me on one of these metaphysical errands, and demonstrate with such ease her facility in a boat and her enjoyment of the peace of just floating in the early morning sun on a lake with a history relevant to our family. That nothing menacing arose from the lake and challenged us was probably a good thing. On the way home, I drove past Energy Lake, a longer, more challenging paddle with a maze of greater complexity hiding Crooked Creek, and she expressed a desire to do some more kayaking with me in the near future. I told her to promise me she'd try Energy in the daytime to look for the crossed trees at the entrance to the swamp. The crossed trees, I thought, would be enough to motivate her.

This would be my last paddle in LBL for a while, and I hoped that taking Libby would create some kind of family legacy. To me, the inland lakes at LBL, while clearly artificial, were conceived within the context of that larger artificial landscape that the high dams, Barkley and Kentucky, had created. The small lakes seemed to exemplify TVA's mission to manage the land in order to preserve its biological diversity, to make it a place where people could go to learn about nature, and to keep it as free from the cacophony of consumer culture as possible. That my father, a working-class man who toiled in a chemical plant most of his life, was able to find a connection to nature in LBL—free of charge, no membership required—seemed a testament to the fulfillment of the best part of TVA's mission. As far as I could see, even though the Forest Service was now in charge, the Land Between the Lakes seemed a place where small dams served a purpose both useful and aesthetically enhancing.

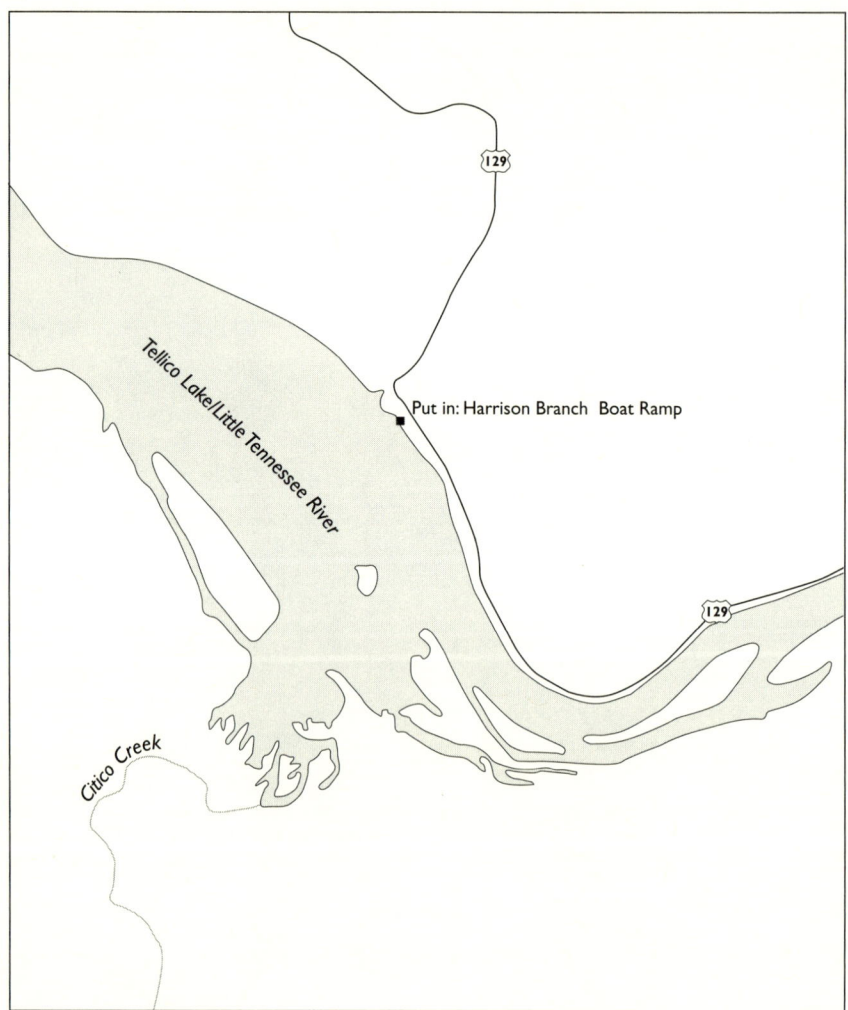

Tellico Lake/Citico Creek

CHAPTER 17

NAVIGATING BY THE STARS UP CITICO CREEK

I was wandering through a Maryville grocery store, lost in the task of gathering the ingredients for a tomato and eggplant casserole, when I came across Drew Crain, a biologist at Maryville College, where I teach. He was staring at the shelf in front of him. He'd harvested a bunch of cucumbers from his garden, he told me, and couldn't find the pickling spices. I glanced at the shelf and spotted the package right away. Sheer luck. Only then, after he was obligated to me for having found the pickling spices, did I propose that he accompany me on a night paddle. He jumped at the idea. Drew is like that, up for anything. I was excited, he was excited, and now all we had to do was pick a time and a place. My first thought was to paddle up Abrams Creek from Chilhowee Lake, which is a segment of the dammed Little Tennessee River, up in the foothills of the Smoky Mountains. This would be a straight shot up a narrow cove to what I remembered as a distinct demarcation of flat water and downhill mountain stream. It would be a safe, easy paddle of a few miles, and I was thinking that we might see or at least hear a bear. The other put-in I was considering was below Chilhowee Dam, on upper Tellico Lake, where we would cross the lake to the mouth of Citico Creek, a maze of passageways that had entertained me on a couple of daylight excursions a few years previous. I'd been searching for two rocks in southwestern Blount County, near Citico Creek, that Donald Davidson, author of a two-volume history of the Tennessee River, described as representing the twin monster hawks of Cherokee myth, Tlanuwa. I didn't find the Tlanuwa but discovered the following: trout circling my boat in clear water; a hornets' nest hanging like a work of art, an amber urnlike shape within the greenery; and the place where the creek began to trickle over gravel into the dammed reservoir. I didn't tell Drew about the Tlanuwa or the hornets' nest, but I did say that finding the mouth of Citico and ascending to its liminal zone would be the more challenging of the two choices.

"Whatever you think," he said. "I'll just follow you."

"No guarantees," I said, thinking that he might have been overly impressed by my discovery of the pickling spices.

A waning gibbous moon was scheduled to rise around 11:30 on the Tuesday night we agreed upon, but I picked up Drew at 9:00, just before dark, loaded his kayak on top of mine, and cinched the plastic hulls down tight. We headed south on curvy Highway 129, traveling toward the Dragon's Tail, a treacherous section that climbed above Chilhowee and Calderwood Lakes, where fairly often motorcyclists went airborne off the mountain, their last ride. Drew broke a silence of a few minutes with this question: "So, Kim, what made you decide to start kayaking at night?"

Drew was the guy who put me up at his rental on the Edisto in the summer of 2010 (see chapter 10), and he knew a bit about my upstream quests. But the tone of his question had a hint of wonder, the same tone that you might have if you asked someone why he liked playing badminton in thunderstorms.

It took me a while to formulate an incomplete answer. I told him it wasn't because I was bored with daytime paddling, not exactly, though in the dog days, when lakes fill up with motorboats and the heat wilts and blurs the landscape, paddling in the cool sanctuary of nighttime had begun to appeal to me. "It's different," I said, the best I could manage at the moment.

We arrived at the Harrison Branch boat ramp two hours before moonrise. Two empty boat trailers indicated that we would have company out there, not really a comforting thought for a kayaker among motorboats. We slanted toward the upstream tip of the island where the Overhill Cherokee village of Citico had thrived as early as the sixteenth century. Now, mostly underwater as a result of Tellico Dam, it would be unrecognizable to Cherokees such as Attakullakulla, also known as Little Carpenter, who traveled to England with Sir Alexander Cuming in the eighteenth century and met King George II; Tsiyu Gansini (Dragging Canoe), his son, who warned that the whites' continued expansion would be "dark and bloody"; or Uskwa'li-gu'ta (Hanging Maw), warrior turned truce maker at this very place in 1782. Now, behind the "Do Not Disturb" warning sign, only buried tools, bones, and spirits dwelled in Citico. The boat ramp's security light cast a narrow white beam across the lake, and we left our headlamps off in an attempt to adapt our vision to starlight.

Drew, who said his wife, Holly, was a bit worried about this night-paddling business, had assured her that here "at least there aren't any alligators."

He then told me that he caught alligators in South Carolina for his dissertation research.

"How do you go about catching an alligator?" I asked.

"Depends on the size," he told me. "If it's a big one, you use a harpoon. Little ones, around four feet long, we'd catch by hand."

A few minutes later, the tree line of Citico village blocked the glow of the security light, and Drew, the guy who used to catch alligators at night in the swamp, said, "This is scary." For a minute I worried that he might have been referring to something specific with the pronoun *this,* as in *"this* whirlpool" or *"this* strange glow" or *"this* beast I cannot identify." By *this,* though, he had meant the strangely disorienting sensation of paddling waters at night. It relieved me now to know that someone as practical and knowledgeable as Drew would have the same feelings I'd been having at night on the water. "Scary," for lack of a better word, described the mixture of thrill and fear that navigating through the darkness elicited.

As we approached one of a few entryways to Citico Creek, skeleton trees and swamp-monster-shaped bushes loomed up ahead. I didn't mention it, but I was thinking of the giant hawks, who according to legend, menaced the Cherokee, kidnapping their children and creating chaos. A medicine man solved the problem. Lowered into their nest by warriors, he threw four baby hawks into the river, instigating a battle between the Tlanuwa and Ukteena, a giant river-dwelling serpent, horned and winged. Ukteena devoured the baby hawks, and the Tlanuwa plucked it from the river and tore it to pieces. They carried it upward, into the sky, and were never heard from again.

Drew gestured vaguely in the direction of an opening that led from the darkness of the lake to the greater darkness of the creek and its encroaching banks. "Does this look right?" he asked. I said I thought that it looked exactly right, and then I admitted I had no idea. It in no way resembled the friendly little passages that I'd seen in the daylight. But we went on ahead into this darker place where the frogs and barred owls raised their voices in protest. Even though it was difficult to make out the banks without a light, you could feel them closing in, sense the change from the open lake to the tighter passages that resembled enclosed pools.

The Cherokee believed in an underworld that warriors entered by way of the still-water pools of rivers and streams; once there, they rode giant rattlesnakes as if they were horses. I didn't know about the underworld, but I was doing my best to avoid going underwater, to stay oriented and upright by taking it slow and keeping my eyes sharp, ears perked. After an hour on the water, I was still feeling a bit shaky. I would be talking to Drew about something and ask him a question, to which he would not respond. Then I'd turn my little lamp on, and he'd be fifty feet ahead, a tiny, faint moving shape, having a conversation with me that I couldn't hear.

When I caught up with him, Drew identified constellations: the Big and Little Dippers, Scorpio, Cygnus the Swan. By locating the North Star, he figured out where the moon would rise, roughly at our backs as we headed into the creek. He told me what kind of frog was making a sound like castanets—a cricket frog—and he distinguished between the two barred owls calling at the periphery of the maze and the great horned owl that started up a warning call farther in.

It was good to know these things. Gradually a glow emerged from the darkness ahead that Drew could not identify. No matter how grounded one is in science and the world of fact, it's difficult not to think of the supernatural, even of alien life, when you see something you can't identify in the woods. We did not speak of these possibilities. Our noses solved the mystery: burning wood. Campfire! We turned on our headlamps, not wanting to startle whoever had found this swamp a fitting place to spend the night. A young man standing near the water's edge told us that he'd just seen a big snake. He was poking around with a stick as if to see if he could bring it up to show us. His girlfriend stood up on high ground, her hands up over her mouth. At what seemed a location too close to the four-foot-high flames, they had a twenty-gallon blue plastic jug of what I guessed was gasoline. Pulled up on the bank was a johnboat with a small outboard motor. The young man said they'd motored across the lake (before dark, of course) and come across this place, a sort of hump of cleared land where people had camped before. He told us that we could go a good ways farther up the creek but that there would be a lot of fallen trees blocking the way. He wished us luck.

We paddled a winding passageway that seemed to go on for miles. Spiders' eyes glowed green from invisible logs. Fallen trees and stumps emerged suddenly in the darkness, our hulls sliding across the submerged ones, tilting us toward oblivion—a good reason to go slowly. Downriver, near the confluence of the Little Tennessee and the Tellico Rivers lived Dakwa, a giant fish that overturned a canoe full of warriors and swallowed one of them whole. The warrior cut his way out with a mussel shell he found in the great fish's belly and was unharmed, except that Dakwa's belly juices scalded the hair off his head, rendering him forever bald. Citico Creek, narrow but thus far surprisingly deep, seemed a good place for Dakwa to arise, nighttime the perfect setting. What you can't see, the imagination fills in, bringing to life the myths of the ancients, those who lived so much closer to the natural world than we do now, we who take an excursion on weekends, perhaps more frequently in the summer, as if nature were some place you had to "go" to, away from the world we have created apart from it.

We came to a place where a fallen sycamore blocked the entire channel, only about twenty feet wide at that point, and we concluded that it was time to turn around.

Back at the campsite, the couple's fire glowed feebly, a faint lamp in the encroaching darkness. They lay beside each other, not touching, staring at the stars, and we slipped past them without a word. Back on the main lake, around midnight, the moon glared orange through a column of clouds, and then cooled to vanilla as it floated into clear sky. Drew let out a yell, startled by a large fish that rolled to the surface right next to him. He remarked that this was "a blast."

"I think that girl was a little freaked out by us coming up the creek like that," he said.

"I don't blame her," I said.

"Something about being out here at night heightens the senses. You really have to pay attention to every move."

Exactly. And this got me thinking about better answers to Drew's question about how I got started night paddling and, more importantly, why I was still doing it. The trip itself seemed answer enough: the sounds of nature, amplified and unfiltered; the infinite depths of the becalmed waters; the cool air, amazingly free of insects; and the way you adapt, after a while, to navigating through the oblivion where imagination goes into overdrive, stirring up whirlpools that contain giant serpents with horns. Yet to relegate the monsters of Cherokee myth to mere imagination seems dismissive, as if myth is irrelevant to the modern world, our enlightened minds dominated by fact and concrete evidence, by infallible numbers and probable forecasts and the absolute certainty with which we deliver our opinions. There were reasons the Cherokee told and retold these stories for centuries. They meant something. The creatures and their battles explained things about nature, which for them commanded reverence and fear, incited respect as well as conflict. Even though a monster like the giant serpent Uktena might seem the embodiment of evil to us, Davidson and others have noted the Cherokees' reverence for snakes. That they created these myths seems a testament to their attempt to explain what can't be seen but felt, those emotions that arise in us at times when we are most sensitive to our surroundings, our senses heightened and tuned in to possibilities beyond what we have experienced in our air-conditioned homes and offices, our backyards and our playing fields.

In her book *Living Stories of the Cherokee,* Barbara Duncan has explained a key principle behind the ceremonies and stories of Cherokee culture: staying in balance, which includes "staying close to the earth and all our relations." We maintain balance by "taking time to dream . . . by recognizing our dark and light sides . . . understanding ourselves and our place in the world around us" (25).

The fallen sycamore blocked us from the literal transitional zone, and we did not feel or see the current of Citico Creek assert itself against the inertia of the lake's pool. Still, finding the mouth of Citico and going up it to what more closely resembled the creek that the Cherokee had lived beside was close enough. In truth, everything beyond the campfire seemed liminal and otherworldly, as if we were somehow poised between the present time of our put-in—at the concrete boat ramp on the dammed lake—and the myths and legends of the past, when human impact was minimal and our efforts more in harmony with the power and mystery of nature. In a sense, we traveled back in time while going from Tellico Lake up Citico Creek, and nighttime intensified that sense of a journey beyond the literal.

Drew was hooked on night paddling, he said, and started trying to think of new places to go. Abrams Creek was still a possibility, and Drew thought that we very well might see bears on that trip, though it would be a long paddle, four miles or so out and back. He remembered a place called Notchy Creek, off the Tellico River arm of Tellico Lake, not far from its confluence with the Little Tennessee River, where Dakwa swallowed the warrior. Perhaps a little fishing would be in order.

CHAPTER 18

WARNING: GERMAN SHEPHERD IN BOW

I was not seeking a replacement for Jasper when Norm arrived: a fully grown but scrawny German shepherd, smelling like desperation and dog pound, his vertebrae protruding like stepping stones. As Julie led him from her car to our front door, I reminded her that I was not ready for another dog. She ignored this rude remark, which I made right there in front of poor Norm. Rescued by Julie from the animal shelter in Jefferson County, where she was working on her mobile spay/neuter job, Norm had been tied to a post in someone's yard for his entire life up until his shelter time, and if she'd left him there, chances were the remainder of his life would have been short. When we brought him into the house, he peed on one white-painted column, and I was thinking, "This is a complication in my life that I could do without." It had been a couple of years since Jasper's death, and I was getting used to the simplicity of not having to worry about a dog, much less a German shepherd, which, as I would find out, required a different level of vigilance than smaller, more ordinary dogs. That first evening Norm didn't seem sure of us either; he was wary and nervous like a coyote. Outside, on the deck, I gave him a bowl of water. He took a long drink, and then he licked my hand. A friendship was formed. Norm had found a new home, and I took back everything I said about him the day I met him, in the fall of 2009.

Two years later, Norm took his place in the bow of my canoe, the same place Jasper occupied on the Tennessee River trip and countless others. He had large paw prints to fill. I tried to avoid comparisons between the two, which placed unrealistic expectations on Norm. Number one, Norm weighs around ninety pounds, twice as much as Jasper, a differential that makes Norm's movements in the canoe of greater consequence than Jasper's. Number two, Norm does not like the water; he will not swim in our pool no matter how we coax him or tempt him with treats and toys, and when we swim, he runs around the perimeter in a panic, barking as if to say, "Get out! Get out! It's dangerous and I do not want to jump in and save you!" Number three, Norm, while intelligent, has his neuroses. He lacks that uncanny ability to remain calm and cool in unfamiliar situations that Jasper had, like when, for example, we were locking through all those dams. Norm, a true force of nature, a pure athlete of a dog, powerful and swift, does not like dogs. Or cats. And when he sees them on land, he goes into perimeter mode, making great lunges and jumping high in the air, whether he's inside our fenced-in lawn or on a leash. It's impressive, reminiscent of dogs with serious jobs: bomb sniffers, bad guy nabbers, dogs that guard banks and jewelry stores. In fact, when people see me walking Norm, they have one of three reactions:

1. They think that I'm some kind of plainclothes cop, Norm my assistant (or the other way around).
2. They fear Norm because of his size and his breed's reputation as a mainstay of the Gestapo during World War II.
3. Or, finally, they say "pretty dog." The word *pretty* is used most often because Norm, for all his faults, has the kind of coloring, the graceful almost effeminate gait, and alert, expressive eyes that would win him some kind of backyard beauty contest, I imagine, if we had one.

In any case, Norm's tendency to freak out around other animals makes him a high-risk first mate in a tippy canoe, particularly with my expensive camera onboard.

Maiden Voyage: Tellico Lake/Baker Creek

I don't know what made me decide to finally get him in the boat after having him for two years, but on a warm winter Sunday morning, just after Thanksgiving 2010, we gave it a try, just to see if it were possible.

"I see a disaster coming," said Julie, who had been telling me to enroll Norm into some kind of doggie behavior school ever since we'd gotten him. I'd been too lazy, and I didn't think Norm would like being told what to do in the company of other dogs. It was somehow beneath him.

We, Norm and I, headed for upper Tellico Lake, to a boat ramp where I felt sure that no other dogs would be around. A dog or two barked faintly from a cabin across the road from the ramp, smoke streaming from its chimney, white against the blue sky. In the ramp lot was one car and trailer, with nobody around to make fun of us (or help) if we turned over or got in a bad fix. Perfect conditions: but when I got him to the water's edge, Norm backed away from the boat with all his might. I even tried pushing him in from behind. He would not get in to retrieve a treat, and he looked at me with grave concern when I sat in the stern and motioned for him to join me.

Finally, I managed to lift him—front half, then rear—into the middle section of the boat, where he stood frozen, his toes splayed, facing the stern, as near as he could get to me without crossing the thwarts. After paddling about ten feet, we were driven back to the ramp by gusts of wind that I had not the strength to battle, Norman tipping half of the boat out of the water, his weight distributed too far to the stern.

I have never seen a dog so happy to leave a boat and return to the car. Off we went down Highway 72 in search of calmer waters. At the Morganton boat ramp, further downstream on Tellico Lake, he jerked the leash out of my hand and trotted just out of my reach the fifty yards back to the car. After much sweet talking and two treats planted in the canoe, we were underway, in water calm enough for me to coax him forward and to get him interested in where we were headed instead of the solid land that we'd just left.

Going up Baker Creek, Norm gazed at the fluttering, clucking pigeons roosting under the highway bridge. He pondered the noonday call of a rooster. When he turned and faced forward, his toes relaxed, his posture more supple and sensitive to the movement of the boat, I sneaked the camera out of my dry bag and silently aimed it at my dog as if he were some exotic species that I would get this one chance to photograph. He turned in profile when I said his name. It was a noble profile. I was thinking, at that moment, how good it was to have company on the water once again. Norm was starting to relax enough to look around and enjoy himself. He hadn't sat down yet, but he wasn't frozen amidships with that worried look on his face. Just as we both started to relax, a test presented itself.

In the distance a hundred yards to starboard, a dog wandered out to the point of a mudflat and waded, sniffing the surface, oblivious to us. Some kind of full-sized long-haired mutt, this was just the type of dog that most enraged

Norm. I put the camera back in the dry bag, readying myself for some serious turbulence. Paddling away from Norm's potential nemesis, I tossed a treat into the bow for distraction.

Norm jumped over the front seat and into the bow so gracefully the boat did not stir. After he consumed the treat and discovered that the confined space would not allow him to swivel his big carcass around to face forward, he looked at me with panic on his dog face. Understandably, Norm, remembering the confinement of his former life, does not like close quarters.

Like a veteran canoe dog, Norm carefully turned and hopped over the seat back to his comfortable position amidships. He stared in the direction of the dog, now sauntering away from the water. I can't say whether he saw him or not, but his demeanor remained calm, as if being in a boat enabled him to rise above such low behavior. He began to enjoy the sun and our silent run to the back of Baker Creek.

After an hour, he lay down. He stood up a couple more times, once to track the flight of a squawking great blue heron, and another time to sniff at a bass boat moored in a pool at the terminus of Baker Creek. He didn't even bark, and when he shifted in the boat, he did it with such grace and care that I never felt we were in danger of capsizing. We were headed upstream, but at winter pool the water was too low for us to get up a creek and through a mudflat to the point where the current would renew itself.

"Looks like you had a great day," said a bass fisherman as we returned to the ramp.

Norm waited for me to tell him it was okay before he disembarked. He hadn't gotten his feet wet all day.

"It was his first time in the boat," I said.

"I never would have guessed it," said the fisherman.

Nor would I, particularly after the first aborted launching. Norm vocalizes constantly with little whines and grumbles and a lovely deep-throated howl that arises when the wail of a siren is within a block or two of our house. If you make a doglike noise or ask him if he wants a treat or to go for a walk, he tilts his head and says something in reply. That day, in the boat, he was quiet the entire time, as if being on the water calmed him. The next test for Norm and me would be an upstream excursion on a river with some current, the Powell, at a place that was completely new to me.

The Powell River

I'd gotten advice from a Forest Service ranger about where I might see the Powell's current kick in again on upper Norris Lake, above Norris Dam. He mentioned, among three or so options, the mouth of Lonesome Creek as a possible put-in for the Powell River transitional zone, near Tazewell. The name itself was enough to motivate me to load up Norm and the canoe. It was mid-April 2011, five months after Norm's first canoe trip.

Never having paddled the Powell, I had no idea what to expect, which is how I usually like it. Bob Lantz, author of *Tennessee Rivers: A Paddler's Guidebook,* called it "one of Tennessee's mystery rivers ... because few have ever been inclined to paddle it and even fewer have produced any record of their floats." I'd paddled a few of the Norris Lake sections in Campbell and Union Counties, but I'd never ventured far enough north to see the actual river that feeds the lake.

To get to the put-in, we descended Lonesome Valley Road, the last seven miles or so unpaved and cratered with deep and frequent potholes. In my years of paddling, I've found that you can measure the beauty of a place by the number of miles of bad road you have to take to get there. We passed under a Norfolk Southern railroad trestle that looked to be sixty or seventy feet high, an imposing structure that seemed to rise out of the middle of nowhere, the bridge itself about all that was visible of the tracks, making me wonder where the trains would come from and where they would go. I could imagine the clackety-clack whooshing sound of a locomotive filling up the narrow valley, the whistle like a summons from the great beyond, far above life down below.

When we got to the gravel lot, the mouth of Lonesome Creek didn't seem so lonesome. A woman and a group of girls fished off the bank, and the lone male, a boy of eight or so, gawked at Norm standing in the back of my Subaru: "That looks like a bear in your car!" he said. Norm, his muzzle poked outside the crack in the window, wagged his tail and whined hello at the boy.

The Powell, swelled from rain, curved between high forested bluffs, just a bit more than a stone's throw across. Flowing into Tennessee at the Hancock County line, the Powell's headwaters are near Wise, Virginia. The river is reputed to be named for an eighteenth-century adventurer who accompanied Dr. Thomas Walker and his expedition into the Cumberland Plateau region; these were perhaps the first white men to set foot in the area. Powell, an early practitioner of self-promotion, got the river and the valley named after him because of how often he carved his name on trees.

Though slow, the current was clearly visible, meaning that I'd put in too far upstream to discover the Powell's liminal zone. I thought I might as well see how far upstream we could get before the rain started or Norm got restless and cramped in the canoe.

The railroad descended from the high bridge and ran alongside the river. Folks were steering four-wheelers down a dirt track, all of this on the right side. The left bank rose straight up sixty to eighty feet, new blooms of dogwoods and redbuds standing out against the pale green of the hardwoods, with an electric smattering of bluebells here and there.

A small fishing boat approached us at a moderate speed. Norm was doing pretty well, but he hadn't lain down and was still a little fidgety, threatening to hop over the thwart and join me in the stern, which probably would have been disastrous. I tossed a couple of treats forward and that kept him busy up in his section of the boat amidships.

When the fisherman got within twenty yards or so, he slowed to idle speed and waved at us. Still, there came our way a little wake, a series of waves such that Norm, in his short career as first mate, had never experienced. Waves or wakes used to drive Jasper crazy; he would reach over the gunnels and bite at the water. Norm stumbled as if drunk when the boat began its side to side rocking, but he did not panic, and soon it was over, and we continued upstream.

After an hour and a half of paddling maybe a mile upstream, Norm and I were both ready to disembark and stretch our legs. Up ahead was the first chance we'd had, a flat green swath of pasture that came down level with the river. Frogs and toads filled up the valley with their bellicose songs.

In the grass where the water seeped, three toads huddled together, one red, the other two brown. Never having seen such a thing, I ordered anxious Norm to stay in the boat while I took pictures with a telephoto lens. Norm waited until I came to my senses and disembarked with my blessing to attend to canine affairs, the toads none of his concern. The toads themselves, oblivious to our landing, went about their springtime business. A bit mystified by what that business was, exactly, I later sent a photo to Drew Crain, my biologist friend and recent night-paddling enthusiast, and asked what these three frogs were up to. Two, I could guess, but three? Drew said they were up to the same thing that most other species were in the spring: mating. The trio comprised two males and one female (American toads), their chances of "success" increased by the two males' joint effort. The female didn't seem to mind.

The toads' rendezvous was at the mouth of a pretty good-sized creek, and since the rain seemed to be holding off and Norm hopped back into the boat without resistance, I headed upstream. On one side was the pasture, beyond it a ridge where the trees were just beginning to bud. On the opposite bank was a flat wooded area where the grassy bank was almost level with the river. The trees were spaced enough so that the grass was even and green, almost like a lawn. I followed the winding creek for a lot farther than I thought it would accommodate us, a half mile at least. Norm stood the entire time, gazing up on the bank and sniffing with his big black nose. I began to wonder what he knew, what he perceived, that I was missing. Probably a multitude of things. I tended to talk to him throughout, voicing whatever came into my head, calming him when he needed it. Now he seemed at peace, his guard down, his eyes narrowed against the sun above the ridge, warming us in that delicious way that it feels after a long winter. I thought there was a lot of hope for Norm as a boat dog of some stature, and I told him so. This helped, I think, once we ran into the creek's transitional zone, a place where the current showed itself against the trunk of a small sycamore, growing midstream. I tried to get past the shoals here to go farther, but the current was too strong for me in the canoe, with Norm aboard, its upstream range much more limited than the kayak's. As we floated back downstream on the Powell past the red buds and the school bus and the railroad tracks, the wind came downriver and began to ruffle the surface, and I let it turn the canoe around backwards. Norm, facing forward in the bow, was bothered by this and said so. When I pointed him back downstream, he snorted and turned his attention to our destination.

Even if Norm might not fully understand the intricacies of the liminal zone—the geography and the science and the abstract spirituality of it—he was enjoying himself, and he seemed to like the calmness of the creek more than the main river, seemed to understand what we were doing as we traveled up it as far as we could go.

Night Paddling with Norm on Calderwood Lake/ The Little Tennessee River

By early fall 2011, I thought Norm was ready for the ultimate test. I'd gotten hooked on night paddling, and I wanted to share this with him, so I picked the relative isolation of Calderwood Lake, where the water lay deep and frigid between the five hundred-foot ridges of Joyce Kilmer-Slickrock Wilderness and the Smokies. Our goal: Slickrock Creek's transitional zone, somewhere upstream from where we would camp, a few hundred yards below Cheoah Dam. This would also be Norm's first camping trip. It was tempting to treat this as another test for Norm, who, it might seem, needed a challenge after passing with high marks on the other voyages. The stakes were much higher here if he freaked out in the boat: the water so cold, the night so dark, the shoreline steep and unaccommodating. But instead of barking orders at Norm like a demented sergeant, as I had been, daring him to mess up, I decided to follow his lead, to let him do what he wanted, to the extent that was feasible, and to see what I could learn from my dog.

The campground was deserted except for a couple of RVs, their generators powering the University of Tennessee football game on a big-screen television. Our site was a half mile away, at the end of the road, the bank almost flush with the water, making it an easy put in. I removed Norm's leash and let him meander. He sniffed the trash in the campfire ring. He ate some grass and gagged. He watered the bushes, the picnic table, and the fire ring. Ate some more grass. Gagged. Finally he lay down with a huff and watched me string my hammock, spread a tarp and an old sleeping bag for him, and unload the canoe.

> **Norm Lesson #1: Let the human do the work.**
> **Lesson #2: If it looks good and tastes good, eat it, even if it makes you gag.**
> **Lesson #3: Stop and smell everything.**
> **Lesson #4: Urinate often.**

It had to be flat dark before we started out in the boat. This was my rule. Waiting for dark, I grew impatient. I worried about Norm capsizing us in the cold water. I worried about the leaves floating downstream, the current generated by nearby Cheoah Dam, a force I'd have to paddle against on the way back to camp. Norm curled up on the tarp and slept while I worried.

Even in the boat, Norm continued this subdued behavior. He lay down on a towel near the middle of the boat and appeared to go to sleep. I tried not to be disappointed, but I felt he owed me some kind of reaction. Was he bored? Did he want me to return to camp after only a few minutes? *Should* I turn around? Did he know something I didn't? I'd heard of horse sense, something my grandfather said I lacked. Was there such a thing as dog sense that humans like me should pay more attention to?

Lesson #5: Relax and lie down in a tippy boat on cold water.

Behind us a waxing gibbous moon rose above the ridge and cast my spindly paddling shadow on Norm's prone body. Pale-gray boulders the size of dumpsters pulsed in the light. The water glistened with sparkly reflection in the wake of my paddle strokes. On the left, where Kilmer's forest grew thick and tall, came the soft, faint roar of a waterfall I *thought* I remembered from paddling here a few years previous. Norm propped his chin on the gunnel and stared toward the ridge top.

A half mile past the waterfall, a beaver plunged his tail twice, and Norm perked up at what I thought must be Slickrock Creek murmuring behind a dark barrier of brush. Inward, toward oblivion, we paddled, away from the moon's eerie light into the unknown. My headlamp revealed a small bank of rocks sluicing a series of rivulets, much too tiny for Slickrock.

We continued following the bend that took us out of the moonglow into the shadow of a ridge. Far above us, small powerful engines whined mosquito-like, bikers throttling through the pass. It was a lonely and forlorn sound, out of place in all of this unlit wilderness. A shooting star streaked across the sky. After two hours of paddling downstream, I asked Norm if it was okay to turn around, even though we had failed to find Slickrock.

On the way back, I turned into the cove where I thought I remembered the waterfall. I couldn't see the waterfall. I couldn't see anything because heavy foliage blocked the moonlight from the narrow passageway and had us groping forward toward a roar that grew louder and louder. I realized that this was not some waterfall I'd seen before. "This is Slickrock Creek," I said to Norm. He glanced my way and reassumed his position, chin on the gunnel, head tilted toward the hidden moon.

Lesson #6: The moon is interesting. Stare at it.

The white noise of shallow moving water filled up the forest and the sky. We ducked under a low limb and came to an opening where a rocky stream poured through a slanting field of rocks twenty yards wide. Patches of moonlight, filtered by the trees, flickered on the moving whitewater. Below the barrier of rocks, we sat in the canoe, the eddy holding us in its spinning embrace.

Back at camp at midnight I started a little fire, and we ate a meager dinner. Despite the successful night run and Norm's impeccable behavior, I thought it best to tether him to my hammock with a six-foot long tie I used for my boat. Norm had been so good I didn't worry about tying an elaborate knot.

I woke up a few times. As the lake rose, my boat made a slapping sound in the small waves ruffled by the night breeze. Falling acorns hit a variety of surfaces: concrete, water, my car. It was a comic campsite to try to get a good night's sleep, but the night was cool, and I felt snug in my sleeping bag, happy that my dog was calm enough to keep still during a night paddle, smart enough to teach me a few things.

Then, just before first light, Norm made me yell at him the only time of the trip. I don't know what it was, but he went after something that should be glad it wasn't caught. He yanked the knot at his collar so hard it came loose, emitted one growl near the campfire pit, and flushed whatever had the temerity to encroach upon our territory. Satisfied, he returned when I called him in a more reasonable tone.

Lesson #7: Forgive humans for yelling at you for things they don't understand.

An hour later, we set out to retrace our route to the creek and back. The morning paddle revealed stark differences between the dammed reservoir and the creek that fed it, the water flashing in the early sunlight, alive like it should be. At night it had talked to us, and now it opened up in all its glory. As we eddied below the tumble of boulders, Norm did not make a sound, but he did stand in the boat, ears up, gazing upstream, as if enraptured. Following his lead, I looked in the same direction, unmoving, trying not to think, until he turned to me, a signal, and we paddled back to camp for the drive home.

Lesson #8: Savor beauty for as long as possible.

CHAPTER 19

FINAL THOUGHTS

> Who looks upon a river in a meditative hour and is not reminded of the flux of all things?
>
> Ralph Waldo Emerson, "Nature"

I returned to the Nantahala River liminal zone in the summer of 2010, three years after that first trip when I entered the cool fog bank, and the carp swam up to my boat in the suddenly clarified water. It was this place that got me started on these quests, the memory of it sustaining my curiosity about other similar places. The disorientation of going from one world to another in just a few paddle strokes, the border between the two so distinctly defined, gave me a sense of traveling from a world of stasis and artificiality to one that was pure and alive. On my return, three years later, I wanted to recapture that wonder, hoping that it would be the same, but knowing, in my heart, that it would not be so. Nothing is.

In what ways, I wondered, would this visit be different? Would my memory of the place distort the possibilities of this second visit? Memory made it impossible not to expect some of the same sights and sounds, some vestige of the original emotions, even though the element of surprise, which made the first trip so remarkable, would probably be absent on this one. At least I thought so.

As on my previous visit, I hiked to the transitional zone before paddling there, starting out at the Nantahala Outdoor Center parking lot, where kayakers and rafts and canoes appeared and disappeared in the roaring whitewater, dipping and charging and rolling in the current. Down the trail beside the makeshift campsites I hiked, next to the diminishing current, and suddenly, in just a few minutes, I'd left the crowds behind. I was alone. At one point, the trail disappeared, and I continued on the railroad track until I reached a point where it veered into the woods away from the river. Below me, at what I thought was the same spot where I'd come upon the willows and the fog bank and the fragments of the bridge, a jet boat sped past two kayakers who bobbed up and down in his wake. No fog, no willows, no bridge ruins. The transitional zone of 2007 was buried underwater.

On the walk back, I came across a woman sitting on the bank, as if studying the river.

"Can I ask you a question about the river?"

She had tattoos on both arms, and she was smiling as she turned away from the water to face me.

"Does the level fluctuate daily?" I asked, hoping that the water level, controlled by a dam upstream, might descend ten feet by morning, when I was set to paddle.

She said it would go down a bit but not much.

I told her what I'd seen three years earlier.

"That was at the end of a ten-year drought," she said.

Having confirmed that the place I sought no longer existed other than in my memory deflated my enthusiasm for waking at 5:30 and being on the water at first light. I did it anyway, telling myself not to expect much and to concentrate on exercise and the joy of being on the water during what I estimated as a six-mile paddle—three miles each way on flat water. Anything could happen, I knew, and even though you might think life predictable, as soon as you became convinced of its predictability and let your guard down, something of interest would rise up. I slogged along in the cool of the morning, and I reached the last bend before the railroad tracks faster than I remembered. Here at the former transitional zone, a lone fishing boat zoomed past me on upriver. The fog alone would have stopped him three years earlier. Now there was nothing to indicate the river's resurrection except some half-submerged trees: big sycamores and oaks, no willows. Where were the willows? Onward I went, up the river, far beyond the place where the current stopped me three years ago.

The waterway narrowed, trees reaching out from both banks, an island up ahead. I passed a tent with a clothesline in front of it, the door open, languid movement within. Somebody lit their first of the day, its blue smoke wafting in the still air. At the tip of the island, the river curved and narrowed more. A brook trickled into it from the left. Up ahead the fishermen who had passed me stood in their bass boat, a father and son, I thought, shrouded in the vapor that rose from the river, the Greater Wesser Falls just above them, a tumble of rocks with water roiling around it. Not a roar now, just a low murmur of water. Later in the day, the release of water from the upstream dam would transform it into a dragon for kayakers to slay. Beyond the fishermen, above the waterfall was the Nanty of lore, the bridges and the shops and the rocks and riffles, all of it distinct in the clarity of the early morning, but deserted now. The fishermen cast their lines in silence, catching nothing, their outlines indistinct, their boat motionless. When I paddled to within yards of them, and they turned and glanced at me, they did not seem real in the fog that rose from the river. And I held my hand out at arm's length and realized that I had entered into the same spell, that I was a ghost myself and that this would become a memory as distinct as the last one, this terminus at the waterfall, a barrier as remarkable as the low fog bank and the clear water I'd discovered three years earlier.

What did I learn from this return? It confirmed some old sayings. You can't step into [kayak] the same river [liminal zone] twice. Nothing stays the same. The hills and rivers endure and will renew themselves, despite our attempts to tame/control/obliterate them. Still, we have a capacity for ruining beauty, for tarnishing it with our engineering and our commercialism. Even if hills and rivers endure and renew themselves, their beauty is fragile, fleeting, vulnerable to manipulation.

Literally I learned that the higher the water of the lake and the river, the farther upstream the transitional zone is pushed. A drought, though devastating to us in so many ways, revealed a place of beauty to me in 2007, though my perception of it as beauty was completely subjective. Floods kill people and destroy property, but they too can reveal beauty and can bring about change that is not completely negative. Going against common sense and the law, Hodding Carter canoed the crest of the Mississippi flood of 2011 for a few days. Except for some coast guard personnel and a few tugs, he and his friends had the river to themselves. Dangerous? Of course. With the Corps of Engineers' wing dams and many of its levees submerged, the river resembled in spirit the same one Mark Twain had navigated in the nineteenth century, the river of old, except that the islands and markers Twain navigated by would be submerged. While Carter has applauded the corps' attempts to engineer protection of the

Mississippi Delta towns and farms, he seems to agree with Andrew Fahlund, vice president of conservation for American Rivers, who said, "We need to give the river room to move," meaning that we should allow some floodplains, well, to flood, to restore them from farms to their natural state so they can help absorb the river's high water. Carter, who grew up near the river, has stated that "While I'd always agreed with the idea of controlling the Mississippi naturally, that was in the hopes of helping the Gulf of Mexico and the wetlands up and down its banks. Now, after experiencing the Mississippi when it was clean-smelling and free, I felt like the river itself deserved a change."

    What I'd seen on a much smaller scale over the past four years made me inclined to agree. What I'd been looking for, hoping for, by paddling upriver to transitional zones was a sense of hope and optimism about the potential of the natural and the pure being aesthetically superior to what we had created with so many high dams. Sometimes the dams and the conditions overwhelmed my attempts to rediscover rivers, as on the James. I felt a bit like Bronwen Dickey, who recently scuba dived to the bottom of South Carolina's Lake Jocassee to explore the vestiges of Mount Carmel Baptist Church cemetery, all that was left of the old community. At 127 feet, near the bottom, where she began to see gravestones, she thought, "This isn't right," and she ascended. She said she'd come to see the damage that was done by the dam, but in the end she realized firsthand what local hunter Dennis Chastain had told her: "Jocassee isn't at the bottom of that lake . . . you could drain it all today and it would never be the same. Jocassee is nowhere, Jocassee is gone." Paddling the wide waters of reservoirs, I got that same feeling of loss, but more often what I found gave me hope that rivers, in some form, will survive our thirsts and appetites, our economic growth, and that, for the most part, rivers and creeks are best left alone, that our experiences on them as paddlers, fishers, boaters, and swimmers are richer when they are alive, in flux, replenishing themselves, despite the economic benefits of reservoirs. Rivers deserve more care and protection, more gentleness than we've given them in the last one hundred years.

    In pursuing liminal zones, I resemble my dog Norman. When we stay at the family farmhouse in western Kentucky, Norman, who is mostly gentle and obedient, cannot stop himself from chasing the cats into the cornfields that surround us, and he won't come back until he's ready, regardless of my yelling, my pleading, my promising of treats. I hear him charging up and down the rows, the leaves shushing at his passing, the tassels stirring. So far, he has not caught a cat, and I hope he never does. Aside from the disillusionment he might feel, it would be awkward for me to explain to the neighbors. In the same way, I hope I never find the perfect liminal zone, for finding it and photographing it will make for one less obsession, one less journey to take on the water in search for beauty, revelation, and mystery. In fact, my quest has not been fueled, I have learned, by a desire for some kind of sublime perfection in the landscape. A sense of mystery, a tension, arises going upstream on rivers and creeks. That curiosity about what might happen around the next bend—beyond the deadfall that rises up to block the horizon, through the obscuring fog, the river like a skilled storyteller who reveals a little at a time, who allows you to ease forward toward discovery—that's what keeps me going out. Rivers and their mysteries don't create happiness in themselves, but they keep me interested, curious, and engaged, and that's the best one can hope for.

EPILOGUE

LETTERS

To My Father

    You would have frowned had I told you where I was going, up Mayfield Creek, from the Mississippi, at Wickliffe, and I would have asked why. Because, you would have said, Wickliffe stinks. My put-in, below a bridge, smells worse than the nearby paper mill. It's a slanting incline of mud slick as ice, ankle deep if you put your weight on it, a hubcap-sized snapping turtle rotting nearby. Don't get snakebit here, you'd say. You're a long way from help.

    The creek took its name from a man who suffered an abrupt shift of fortunes. Flush with winnings at a race in Hickman, Kentucky, Mr. Mayfield was robbed and kidnapped. In his attempt to escape, he was shot and fell dying into the creek. At some point he had carved his name on a tree.

    Good place to get knocked in the head, you would say.

    I've got your fly rod, the poppers you made by hand, with a new leader I bought yesterday from an outdoors store that opened long after you died. You would have liked it there. You'd like to see me catch a fish, I know, at least to try. But now I'm shooting pictures: ducks, a heron, bridges, the green sunlit scum I've broken through, everything the opaque water reflects. It doesn't look so good for fishing.

    Why are you doing this? you would ask.

    Mayfield's bones lie below me, buried in the bottom mud. Little did he know, as he watched his lifeblood mingle with the slow, muddy current, that his murder would become a martyrdom, that somebody carried his name forward, repeated it until his death begat the creek's identity, the carving in the tree his last words, in effect, his name containing the verb for permission—*may*—and the second part—*field*—resonant of natural beauty, together forming a convergence of optimism and possibility.

Mayfield Creek is straight like a canal, dark and quiet, as if subterranean. Long, lean, and toothy, gars swirl the water and bump against my hull. Something grabs my paddle and lets go when I yell. I learned by watching you, one arm holding the wooden paddle near the blade, steering from the stern, casting with the other arm to within inches of the bank. You'd untangle the lures I overcast from the bushes as willow flies swarmed, having to plunge your hands through clusters of leaves that harbored, I imagined, a host of varmints poised to strike, bite, or pinch. Sure enough, you were right to warn me: a snake lies on a fallen sycamore in a splotch of sun not five feet from a narrow passage, laid out among the leaves, yellow and brown like him. It's as if he's there to extract a toll. You've killed a snake or two, but you'd leave him be like I did.

The creek is twenty yards wide now, open to the sky, the sun strengthening to more than mere insinuation. I squint. Up ahead the deadfall closes in, and the smell changes as the current charges through a logjam. Had he been one step faster, the log wider and more stable, Mayfield might have crossed here to safety. When we fished, we stayed clear of places like this, too narrow for your riveted-steel fishing boat, the Larson, made in Little Falls, Minnesota, propelled by the twenty-five-horsepower Evinrude. You kept me out of the main channels where the barges ran, and we never once sampled the Big Muddy, not that far away really. You let me discover those places on my own, in my own way.

On the way back, the fly rod snaps the new leader on the second cast. As I paddle toward the unmoored popper, the small cork you painted black, sprouting the yellow plumes you made from a buck's tail, as I lean over and put my hand in the water, I think of gar, copperhead, whatever grabbed my paddle a ways back. Mayfield's bones. Why is it I'm doing this? I'll tell you when I know, though let's just say ghosts live in rivers, and it's only in the narrow places, the shallow channels that the snakes guard, where I paddle with half strokes, in solitude, that I realize the depths that we reached in silence, sharing the same boat, the same tackle box, having arisen from the same land and rivers.

To Clarks River

The railroad bridge spans a shorter distance than I recall, my sprint across it epic in memory, the spaces between the ties gaping chasms, my legs heavy, as if filled with sand, my breath ragged with panic. Drought has stunted your flow, the deep, blue pools of memory now shallow and turgid, flesh and bones of the blue gill we fought buried and desiccated into fragrant bottom mud too soft to walk upon. Along your banks we cut and lit grapevines that you grew, sweet smoke that burned the throats of friends I rarely see: the chubby rich kid with the contagious chuckle, and the small quick athletic one who stuttered. Beside the stands of tall green cane on gravel bars we speculated about what sex would be like with a select group of seventh-grade girls and a student teacher that the rich kid swore he'd seen modeling underwear in a catalogue. Between the lush corn and your clear blue pools, we coughed and chuckled our way through the nights and woke in sleeping bags damp with dew, our jaws aching from laughter. The rich kid, twice married now, no longer chuckles, no longer seems so wealthy, and the small one, who owns a furniture store, lost his stutter and the inclination to dodge and fake with a football.

You intersect my life at intervals in your course from Calloway County to the Tennessee River, a hundred crooked miles that crossed and recrossed the railroad tracks where we lay coins for trains to flatten. You nurtured hoboes and wild dogs, which we feared, the dogs a line of hazy dots always running toward us down the tracks from miles away, the hobos myths our parents told to keep us from you. Two counties over, wider and deeper, you pour my memories into the Tennessee. Here you pass below a highway bridge, an empty boat ramp decorated with the usual: potato chips have been chewed and swallowed, the bags bright red against the gray concrete, night crawlers from blue plastic tubs skewered and submerged until they stopped wriggling and hung limply in your dark depths. You play host to a lingerie/sex shop, the "ultimate," it claims on concrete block painted lavender, a railed deck facing the water.

What have you seen here, river? Love and meanness in cheap lace and nylon? Blood splashed on the riprap that suffocates your bank? What memories took their root here in the thick mud that smells like turtle? A thin man in the bow of a tilted boat tells me he's here every morning, that the river is up six feet and the fish won't bite. That back in April he saw roosting woodpeckers high in the bald cypress and sycamores. You can go all the way to Benton, says the man, maybe even farther in that thing. I paddle under two bridges, one railroad, one number 284, old Benton Highway, short way by land. If only I'd had this boat years ago, I would have known the length of you, not just the section beside the railroad tracks. Now, going upriver as far as you will tolerate, I'm trying to reclaim the miles, to fabricate memory too deep to retrieve. I'm a fish bumping bait against the bottom, blinded by muddy water. Deer huff at my approach, as I glide, paddle raised. Gar lollygag, impudent as they expose their spotted tails. It looks as if they've smeared themselves with mud, these lords of the river, all day long wallowing, bubbling your surface. One chases the popper I cast, curious, and I feign to snag him. And what would you eliminate, river, if you could? The gar? The carp? The mosquito? The human? His dams? His drains? His waste? This looks like a good place to stop and rest, here below the static of Interstate 24, between the giant concrete piers, here in the weeds where nothing but your current stirs. You undercut the traffic above, your lanes empty of human commerce, empty of frenzy and panic, populated only by boatmen like me and the thin man who have lost their way on land, who search for memories in currents whose headwaters are clogged with deadfall. Speak to me. You know something, river, that I can't quite fathom. You'll tell me if I'm quiet and still, if I linger and study your run from my birthplace under the railroad tracks, a mere creek that we could jump across, through Benton and Calvert City, widening and gathering strength, under Interstate 24, past the lingerie shop, the barge terminal, slowing down to mingle with the Tennessee, to veer and ride the Ohio, and then to be taken up by the implacable Mississippi and carried to the Gulf of Mexico.

BIBLIOGRAPHY

Abbey, Edward. *Down the River.* New York: Plume, 1991.

Ambrose, Stephen. *Undaunted Courage.* New York: Simon and Schuster, 1997.

Branson Landing. "The Fountains." http://www.bransonlanding.com/fountains.html.

Carter, Hodding W. "57 Feet and Rising." *Outside* (August 2011): 72–79, 108, 115.

Confederated Tribes of Warm Springs. "History and Culture: Chronology." <http://www.warmsprings.com/Warmsprings/Tribal_Community/History_Culture/Chronology/>.

Conrad, Joseph. *Heart of Darkness.* 1903; New York: W. W. Norton, 1988.

Creason, Joe. "Playground for the Rugged." *Louisville Courier-Journal,* June 22, 1963. <http://www.imnothere.org/DocPage.htm>.

Davidson, Donald. *The Tennessee.* Vol 1. Knoxville: University of Tennessee Press, 1978.

Dickey, Bronwen. "The Rapture of the Deep: Diving the Sunken South." *Oxford American* (Best of the South 2011): 44–48.

Dorward, Frances Brown. *Dam Greed.* Philadelphia: Xlibris, 1991.

Driggers, Ken, and Bill Price. *Edisto River Companion.* Columbia, SC: Palmetto Conservation Foundation, 2008.

Emerson, Ralph Waldo. "Nature." In *Ralph Waldo Emerson: Selected Prose and Poetry.* New York: Holt, Rinehart, and Winston, 1966.

Foote, Shelby. *The Civil War.* Vol. 2. New York: Random House, 1963.

Garmire, Sean. "Seized Pot Worth 25M to 60M." *Eureka (CA) Times-Standard.* June 26, 2008.

Golze, Alfred R., ed. *The Handbook of Dam Engineering.* New York: Van Nostrand Reinhold, 1997.

Hemingway, Ernest. "Three Shots." In *The Nick Adams Stories.* New York: Scribner's, 1981, 3–5.

"How Tlanuwa Defeated Uktena (Cherokee)." Indians of Arkansas website. <http://arkarcheology.uark.edu/indiansofarkansas/index.html?pageName=How%20Tlanu wa%20Deafeated%20Uktena%20%28Cherokee%29>.

Lane, Belden. *Landscapes of the Sacred*. Baltimore: Johns Hopkins University Press, 2001.

Lantz, Bob, and Bob Sehlinger. *Tennessee Rivers*. Knoxville: University of Tennessee Press, 1979.

Least Heat-Moon, William. *Blue Highways*. Boston: Little, Brown, 1999.

Mazel, David. "Annie Dillard and the Book of Job." In Annie Merrill Ingram et al. *Coming into Contact: Explorations in Ecocritical Theory and Practice*. Athens: University of Georgia Press, 1997, 185–96.

Maclean, Norman. *A River Runs Through It and Other Stories*. Chicago: University of Chicago Press, 1976.

———. *Young Men and Fire*. Chicago: University of Chicago Press, 1993.

McCulley, Patrick. *Silenced Rivers: The Ecology and Politics of Large Dams*. New York: Zed Books, 2001.

Medred, Craig. "When Fear Overpowers Reason, Then You're Lost." *Anchorage Daily News*, June 29, 2008.

Neely, Jack. "Tellico Dam Revisited." *Metro Pulse*, Dec. 9, 2004.

Nute, Grace Lee. *The Voyageur*. St. Paul: Minnesota Historical Society Press, 1955.

Raban, Jonathan. *Driving Home: An American Journey*. New York: Pantheon, 2010.

Rennick, Robert M. *Kentucky Place Names*. Lexington: University Press of Kentucky, 1988.

Slaven, Janie. "Yahoo Falls Monument to Be Removed." *McCreary County Record*, Sept. 12, 2007. <http://mccrearyrecord.com/local/x154969040/Yahoo-Falls-monument-to-be- removed>.

Smith, Frank E. *Land Between the Lakes: Experiment in Recreation*. Lexington: University of Kentucky Press, 1971.

Stepp, Joe. "Notes on Aquone." Nantahala, North Carolina: Land of the Noonday Sun. <http://www.nantahalanc.com/NotesByJoeStepp.html>.

Swift, Jonathan. *Gulliver's Travels*. 1726; New York: Penguin Classics, 2011.

Swiman, Elizabeth, et al. "Living with Alligators: A Florida Reality." University of Florida IFAS Extension. <http://edis.ifas.ufl.edu/uw230>.

Tankersley, Kenneth Barnett. "Yahoo Falls Massacre, McCreary County, Kentucky." <http://freepages.genealogy.rootsweb.ancestry.com/~brockfamily/YahooFalls-byKTankersley.html>.

Tennessee Valley Authority. *Final Environmental Impact Statement on the Natural Resource Management Plan at Land Between the Lakes,* Oct 1994, Vol 1 QH 105T2F55 1994.

Thompson, Jim. *Tellico Dam and the Snail Darter.* Knoxville, TN: Spectrum Communications, 1991.

Twain, Mark. *Adventures of Huckleberry Finn.* London: Chatto & Windus, 1884; New York: Charles L. Webster, 1885.

———. *Roughing It.* Hartford, CT: American Publishing Company, 1881.

Washington, Margaret. "Gullah Attitudes toward Life and Death." In Joseph E. Holloway, ed., *Africanisms in American Culture.* 2nd ed. Bloomington: Indiana University Press, 2005, 152–86.

World Wide Fund for Nature. "Free Flowing Rivers: Economic Luxury or Economic Necessity?" *assets.panda.org/downloads/freeflowingriversreport.pdf.*

E 169 .Z83 T74 2013
Trevathan, Kim, 1958-
Liminal zones

AUG 0 6 2013